# Max Axiom and the Society of Super Scientists

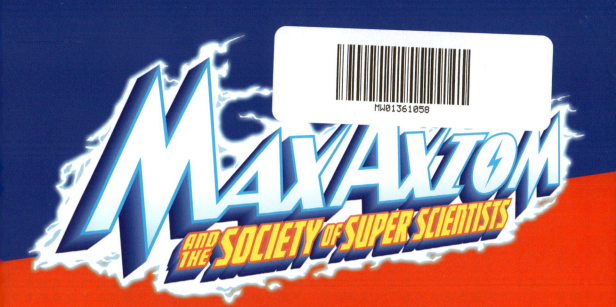

## Defending the Earth

by Carol Kim
illustrated by Daniel Pedrosa

Capstone Press
a capstone imprint

Published by Capstone Press, an imprint of Capstone.
1710 Roe Crest Drive North Mankato, Minnesota 56003
capstonepub.com

Copyright © 2023 by Capstone. All rights reserved. No part of this publication may be reproduced in whole or in part, or stored in a retrieval system, or transmitted in any form or by any means, electronic, mechanical, photocopying, recording, or otherwise, without written permission of the publisher.

Library of Congress Cataloging-in-Publication Data
is available on the Library of Congress website.
ISBN: 9781666337297 (hardcover)
ISBN: 9781666337303 (paperback)
ISBN: 9781666337310 (ebook PDF)

Summary: Most objects that fall to Earth from space are harmless. But occasionally much larger objects pose a real threat. If a large asteroid or comet hits Earth, it could change life as we know it. Is there a way to stop these threats before it's too late? Ride along with Max Axiom and the Society of Super Scientists and learn about Near-Earth Objects and ways scientists think we can defend our planet.

Editorial Credits
Editor: Aaron Sautter; Designer: Brann Garvey; Media Researcher: Morgan Walters; Production Specialist: Polly Fisher

All internet sites appearing in back matter were available and accurate when this book was sent to press.

# TABLE OF CONTENTS



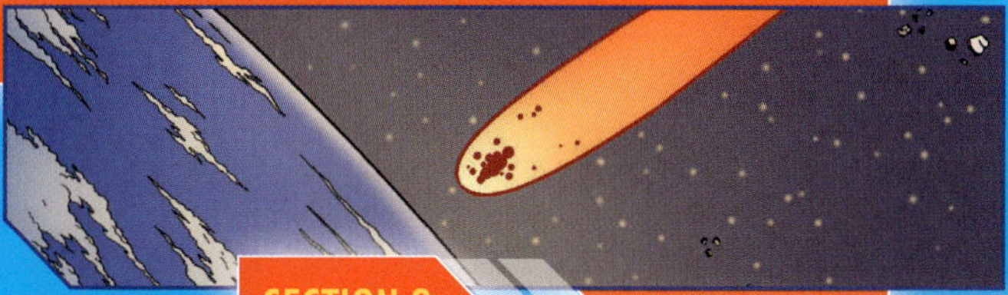



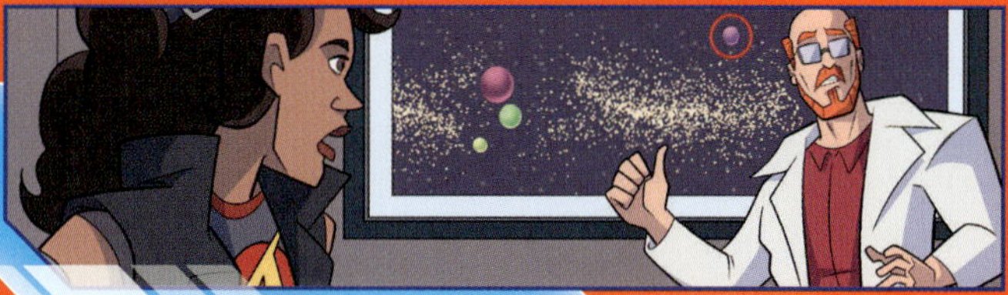



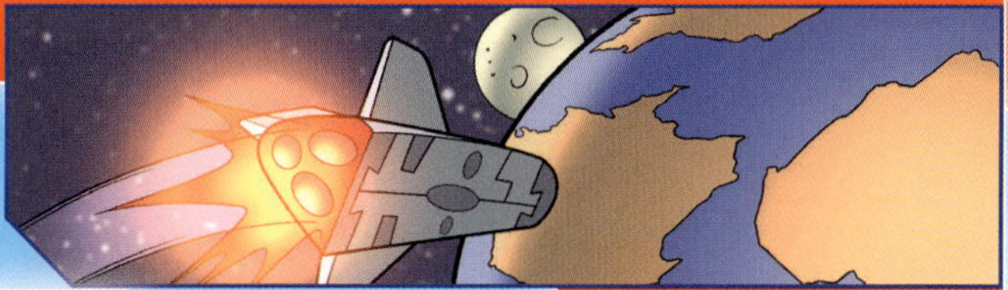



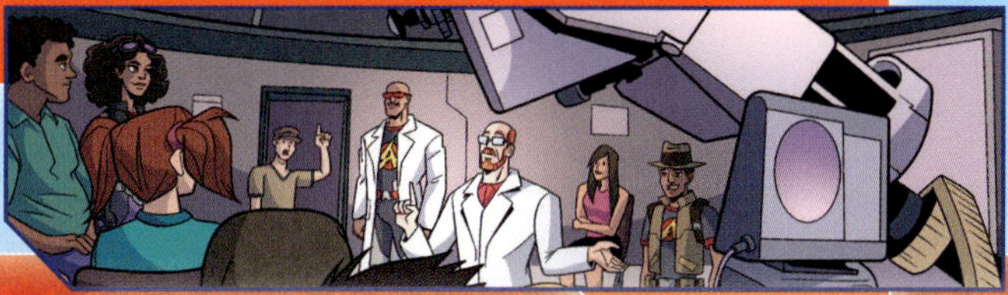

# THE SOCIETY OF SUPER SCIENTISTS

## MAX AXIOM

After years of study, Max Axiom, the world's first Super Scientist, knew the mysteries of the universe were too vast for one person alone to uncover. So Max created the Society of Super Scientists! Using their superpowers and super-smarts, this talented group investigates today's most urgent scientific and environmental issues and learns about actions everyone can take to solve them.

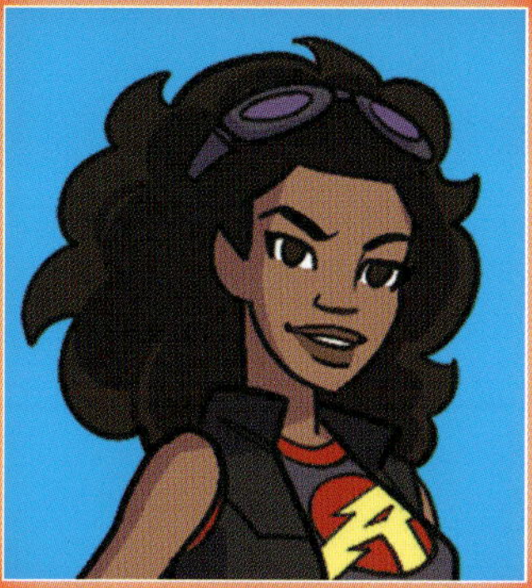

LIZZY AXIOM

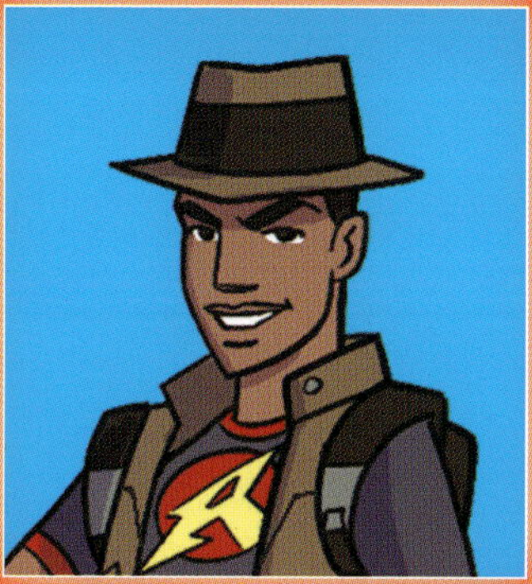

NICK AXIOM

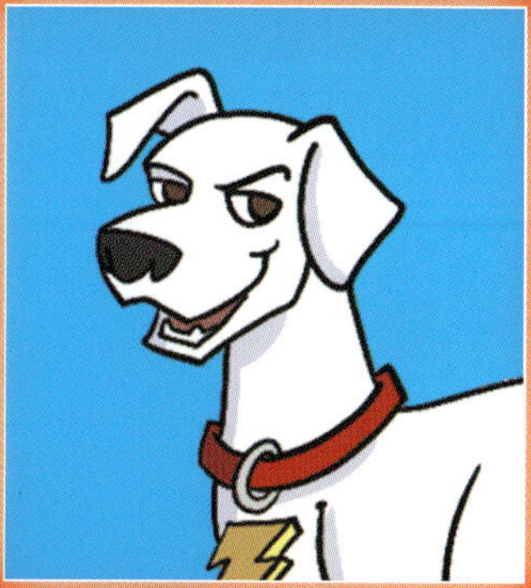

SPARK

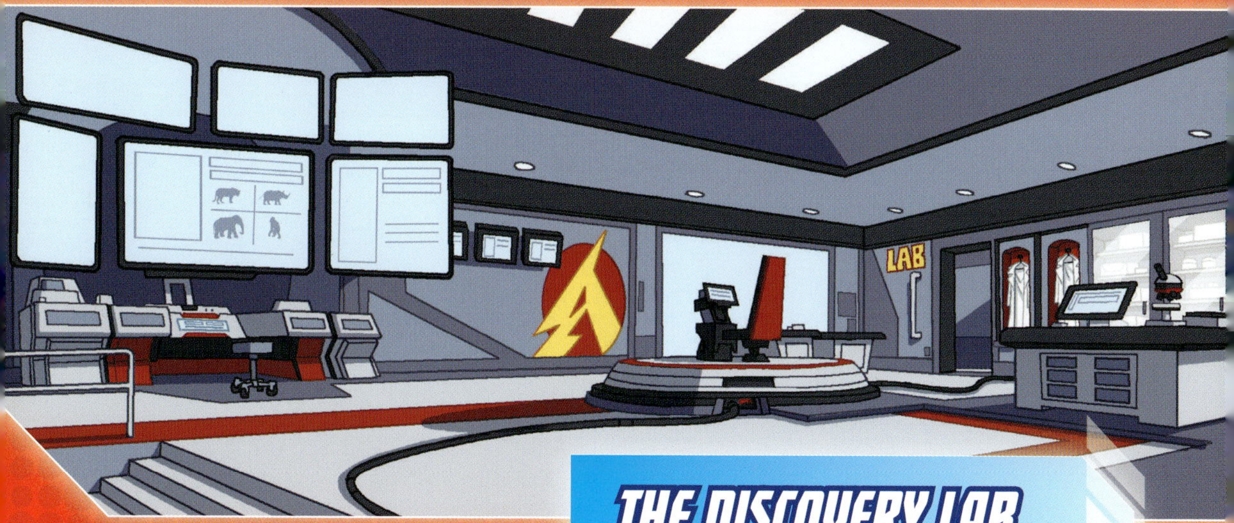
LAB

## THE DISCOVERY LAB

Home of the Society of Super Scientists, this state-of-the-art lab houses advanced tools for cutting-edge research and radical scientific innovation. More importantly, it is a space for the Super Scientists to collaborate and share knowledge as they work together to tackle any challenge.

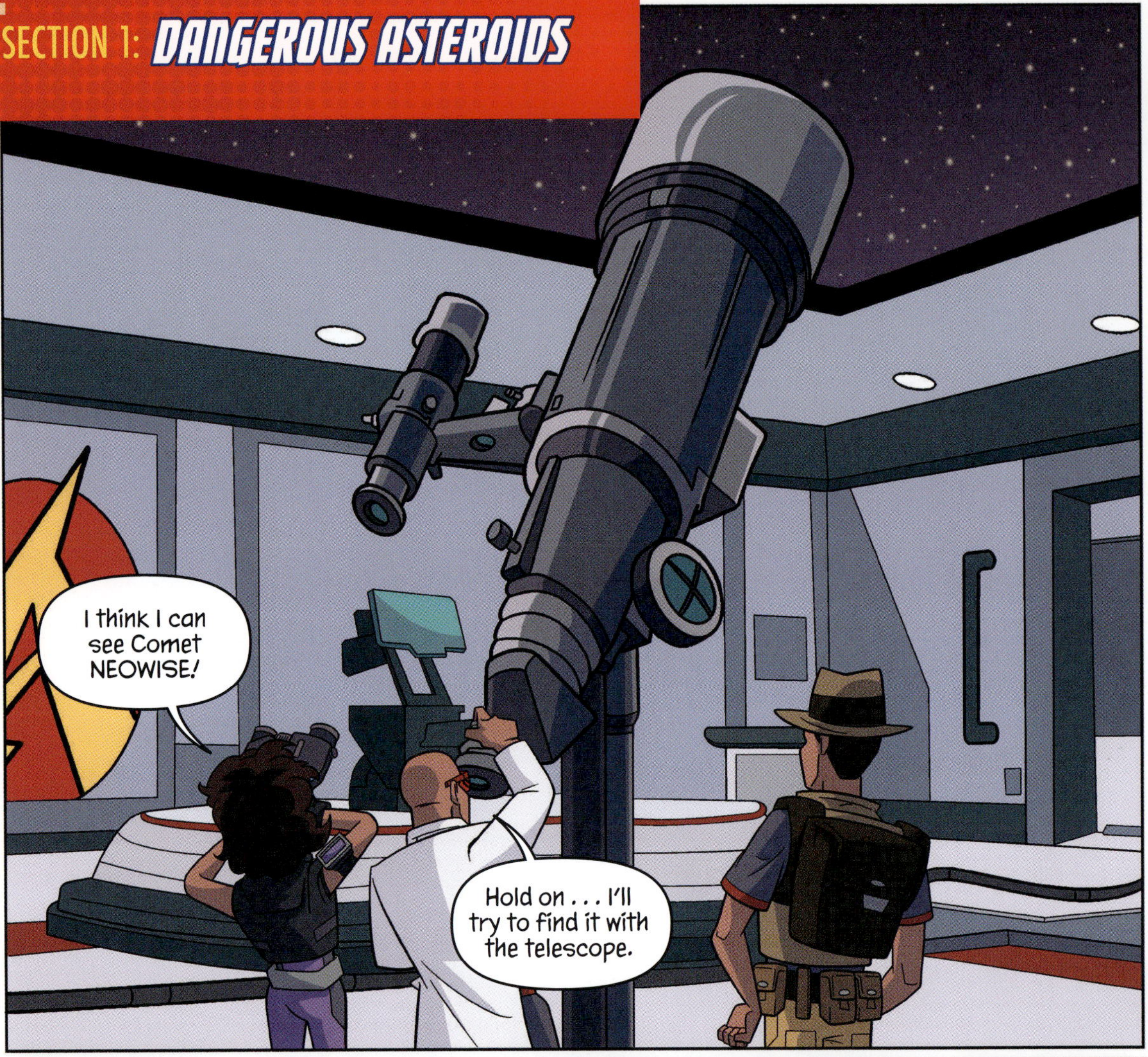
I think I can see Comet NEOWISE!
Hold on . . . I'll try to find it with the telescope.

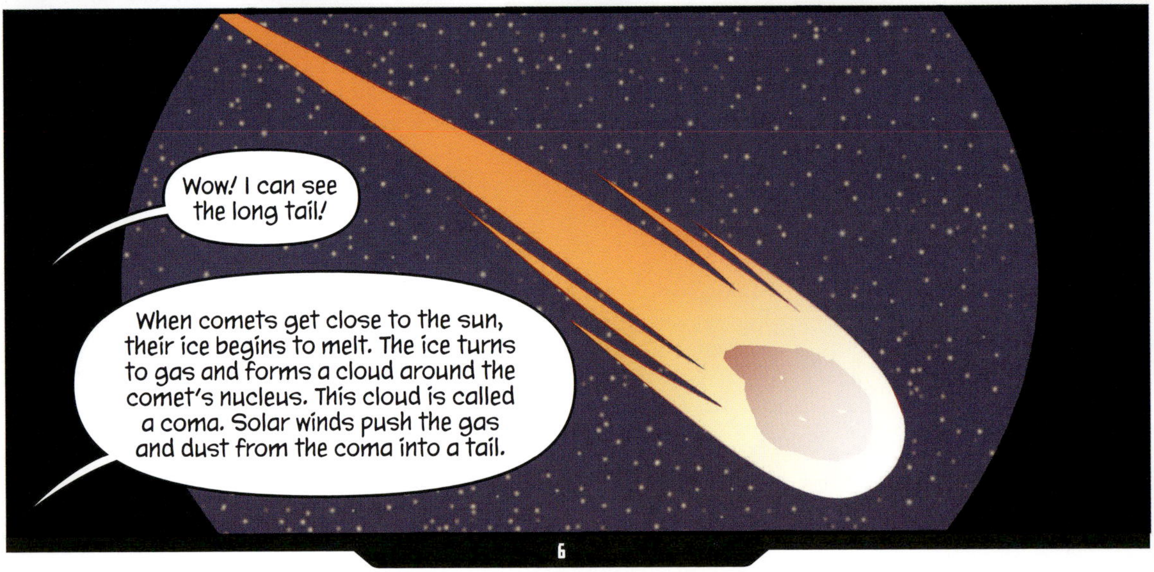
Wow! I can see the long tail!
When comets get close to the sun, their ice begins to melt. The ice turns to gas and forms a cloud around the comet's nucleus. This cloud is called a coma. Solar winds push the gas and dust from the coma into a tail.

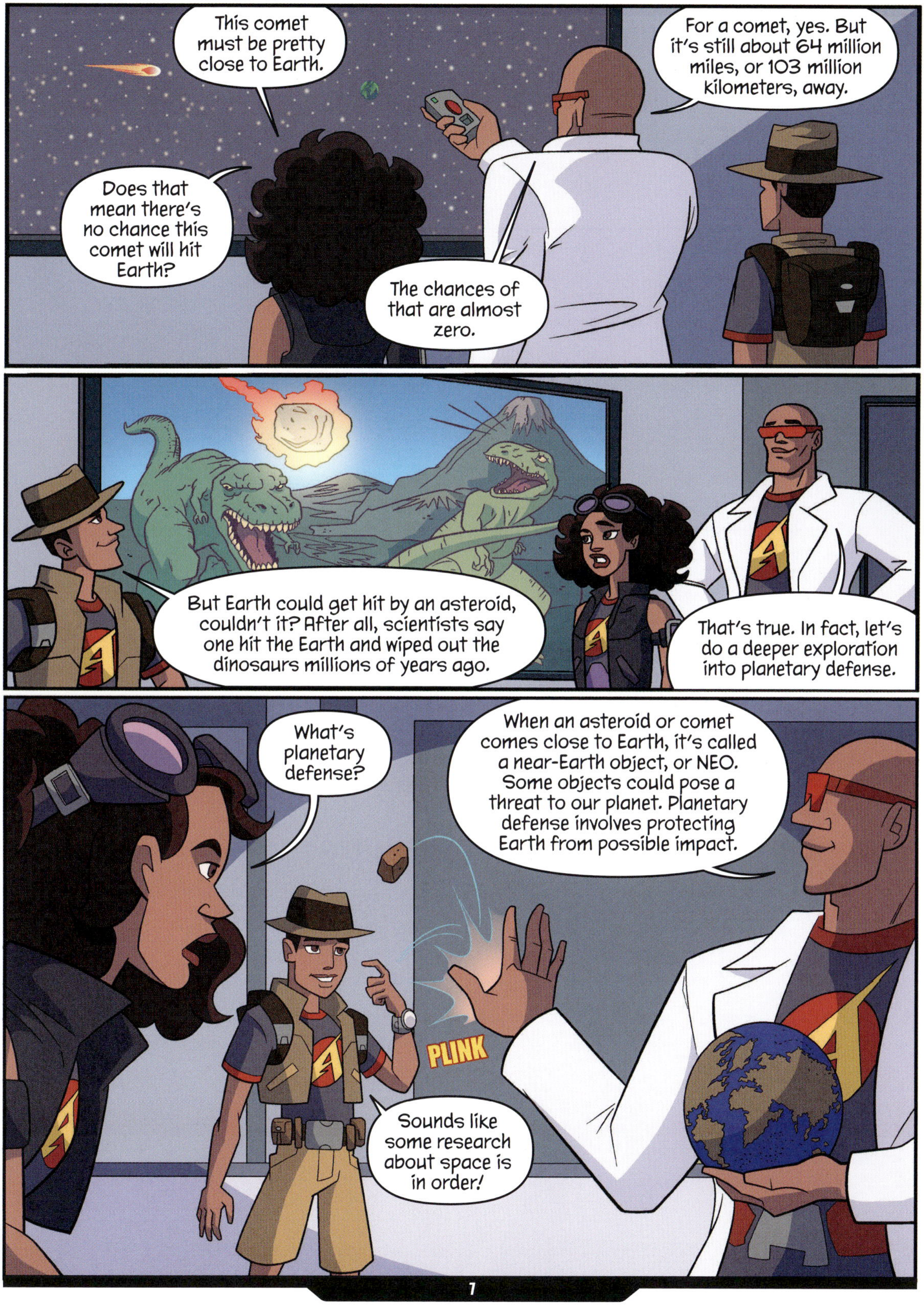
This comet must be pretty close to Earth.
For a comet, yes. But it's still about 64 million miles, or 103 million kilometers, away.
Does that mean there's no chance this comet will hit Earth?
The chances of that are almost zero.
But Earth could get hit by an asteroid, couldn't it? After all, scientists say one hit the Earth and wiped out the dinosaurs millions of years ago.
That's true. In fact, let's do a deeper exploration into planetary defense.
What's planetary defense?
When an asteroid or comet comes close to Earth, it's called a near-Earth object, or NEO. Some objects could pose a threat to our planet. Planetary defense involves protecting Earth from possible impact.
PLINK
Sounds like some research about space is in order!

What are we looking at?
This is the asteroid belt. After the solar system formed, some matter didn't become part of the sun or planets. This matter continues to orbit the sun, mostly between Mars and Jupiter.
Asteroids are usually made up of rock or metal. Comets are made up of ice, dust, and rocks.
That's right. Asteroids can range in size from a few feet to hundreds of miles wide. The nucleus of most comets tends to be about 6 miles, or 10 kilometers wide.
What about meteors?
Meteoroids are small pieces of rock or ice. When they enter Earth's atmosphere, they're called meteors. Most burn up as they fall. But if they crash into Earth, they're called meteorites.
Scientists believe NEOs larger than a half mile, or 1 kilometer, would have devastating effects on Earth.
That would be a little more than two U.S. Empire State buildings. But asteroids that size come around only every few million years.

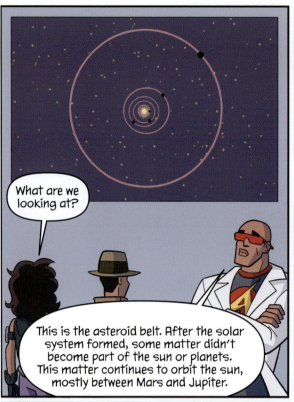

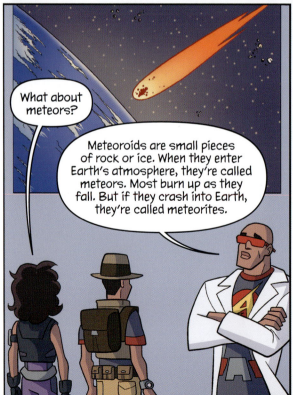

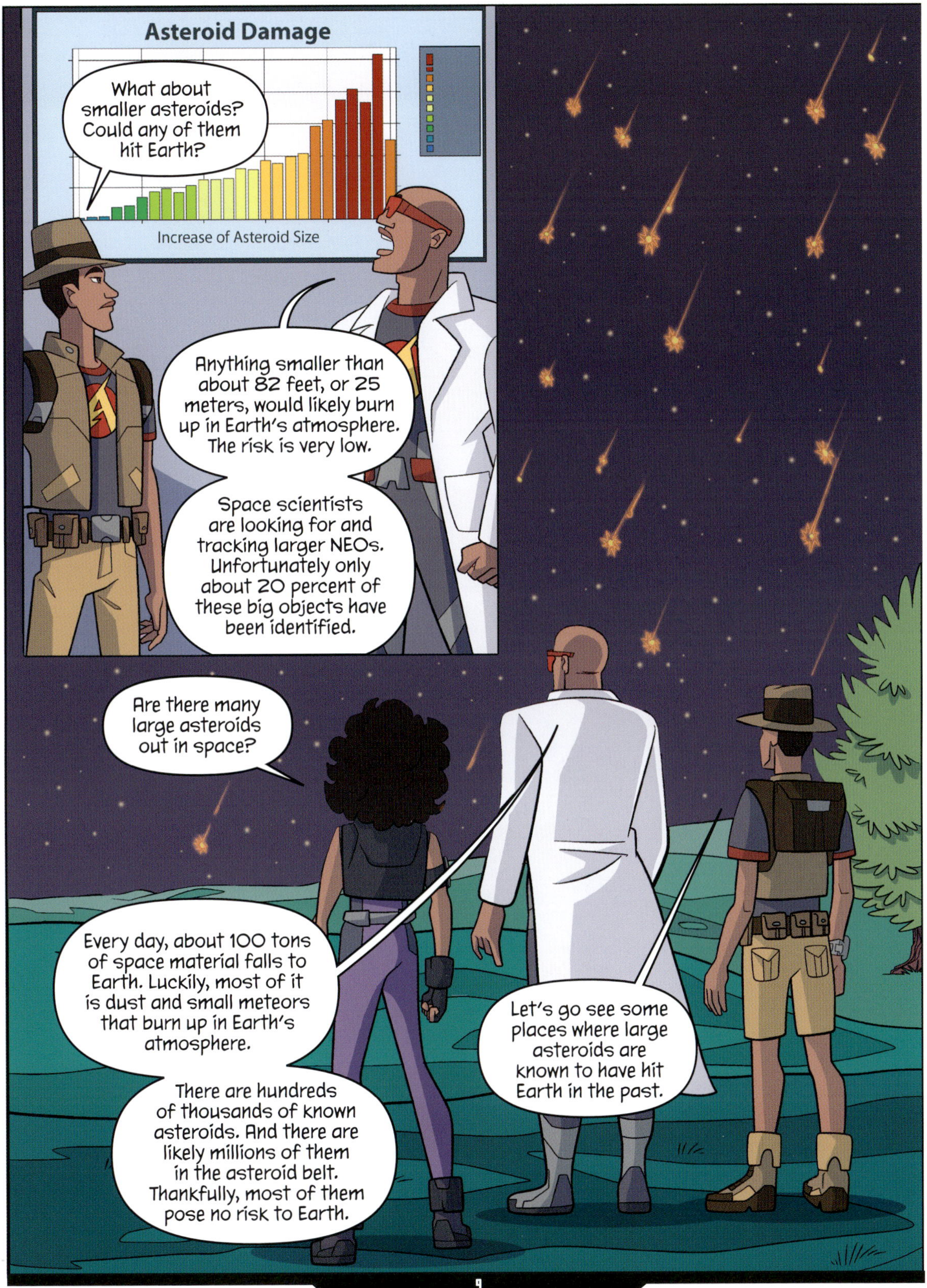
Asteroid Damage
Increase of Asteroid Size
What about smaller asteroids? Could any of them hit Earth?
Anything smaller than about 82 feet, or 25 meters, would likely burn up in Earth's atmosphere. The risk is very low.
Space scientists are looking for and tracking larger NEOs. Unfortunately only about 20 percent of these big objects have been identified.
Are there many large asteroids out in space?
Every day, about 100 tons of space material falls to Earth. Luckily, most of it is dust and small meteors that burn up in Earth's atmosphere.
There are hundreds of thousands of known asteroids. And there are likely millions of them in the asteroid belt. Thankfully, most of them pose no risk to Earth.
Let's go see some places where large asteroids are known to have hit Earth in the past.

The most famous asteroid to hit Earth was the one that wiped out the dinosaurs about 66 million years ago.
The asteroid was about 7 miles, or 11 kilometers, wide. The crater it left behind is located mostly underwater in Mexico's Yucatán peninsula.
Below is the Barringer Crater left by a large asteroid that struck Earth 50,000 years ago. It's located near Winslow, Arizona.
KRAAAKOOOOOMMM!
Now let's travel to 1908 in Tunguska, Siberia in Russia. An asteroid about 164 feet, or 50 meters, wide exploded here. It didn't actually hit the ground
Instead, it exploded in mid-air. The blast flattened 80 million trees over an area measuring 830 square miles, or 2,150 square kilometers.
KABOOOOM!
Welcome to Chelyabinsk, Russia in 2013. Look to the skyline and you'll see the trail of an asteroid that exploded near the city. The blast damaged thousands of buildings. Many people were injured by flying glass from shattered windows.

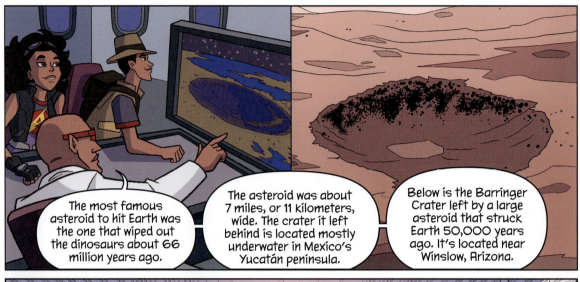

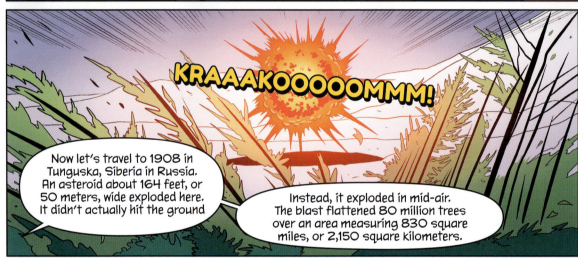

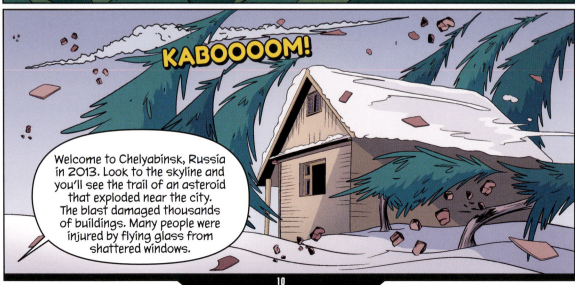

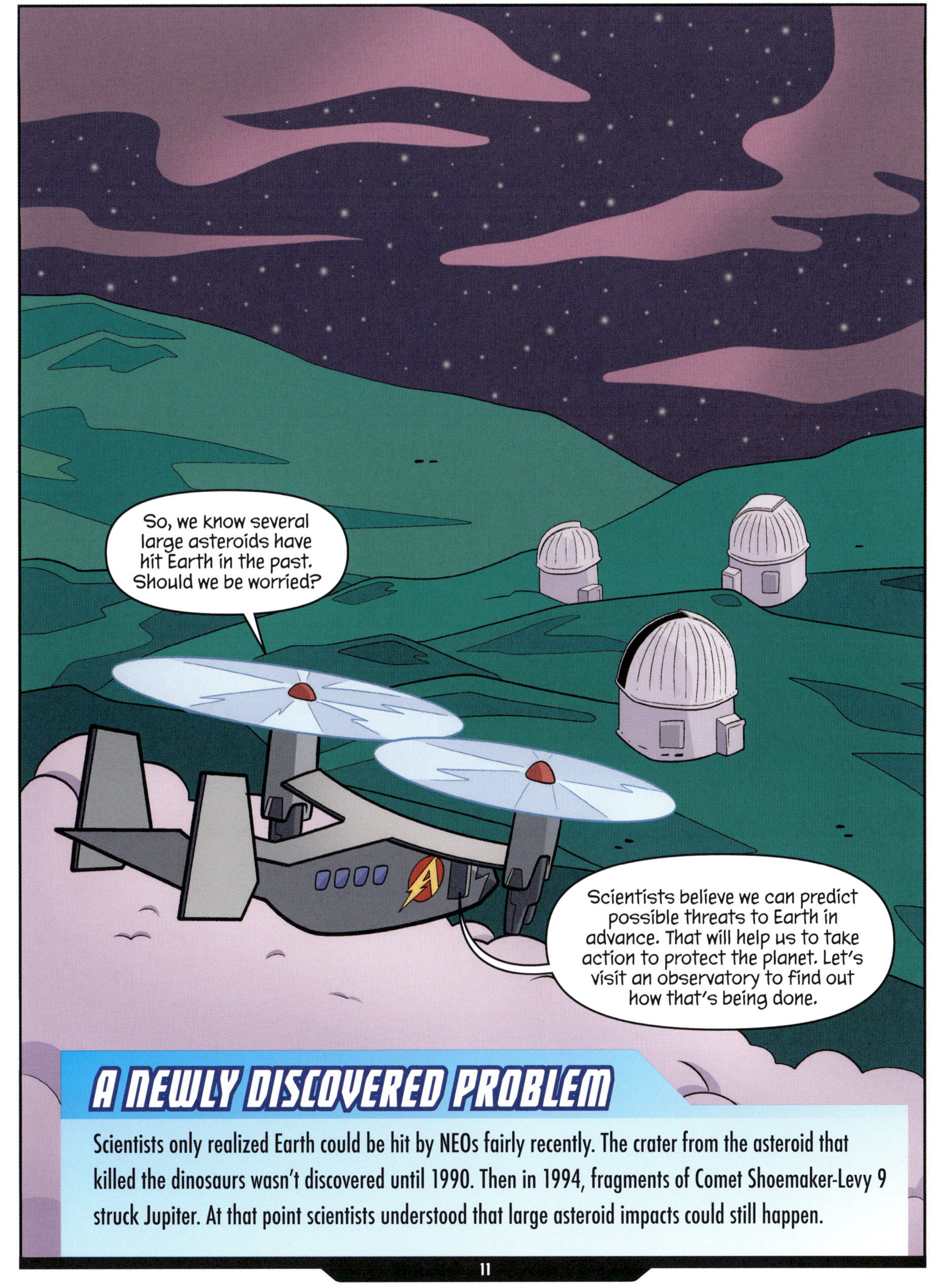
So, we know several large asteroids have hit Earth in the past. Should we be worried?
Scientists believe we can predict possible threats to Earth in advance. That will help us to take action to protect the planet. Let's visit an observatory to find out how that's being done.

## A NEWLY DISCOVERED PROBLEM

Scientists only realized Earth could be hit by NEOs fairly recently. The crater from the asteroid that killed the dinosaurs wasn't discovered until 1990. Then in 1994, fragments of Comet Shoemaker-Levy 9 struck Jupiter. At that point scientists understood that large asteroid impacts could still happen.

# SECTION 2: PREVENTION IS PROTECTION

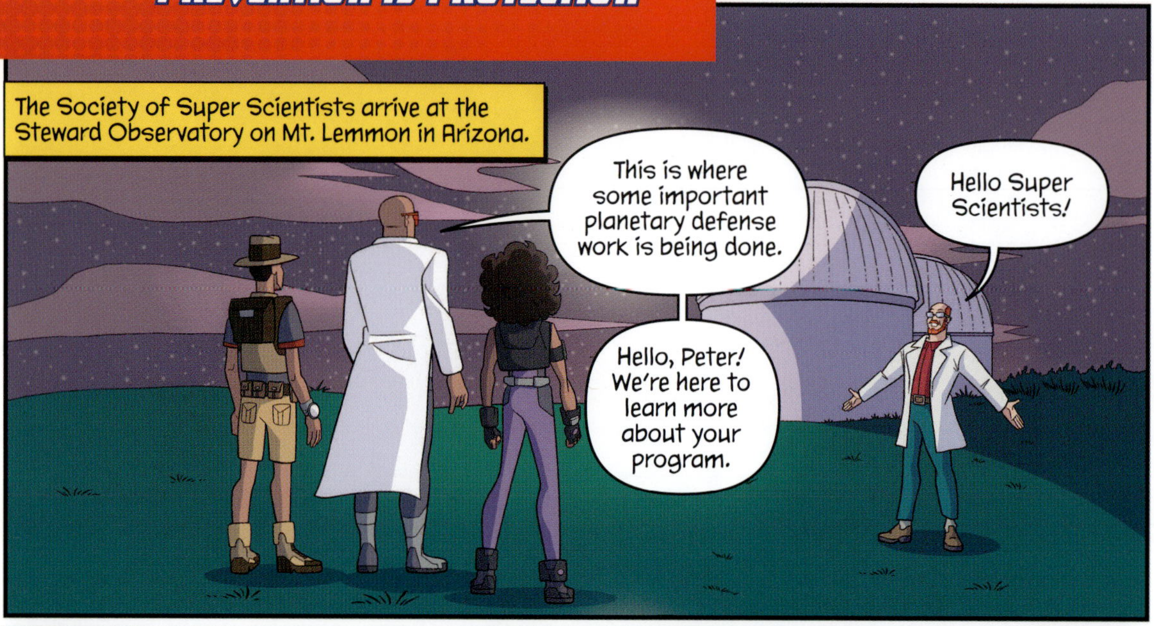
The Society of Super Scientists arrive at the Steward Observatory on Mt. Lemmon in Arizona.
This is where some important planetary defense work is being done.
Hello Super Scientists!
Hello, Peter! We're here to learn more about your program.

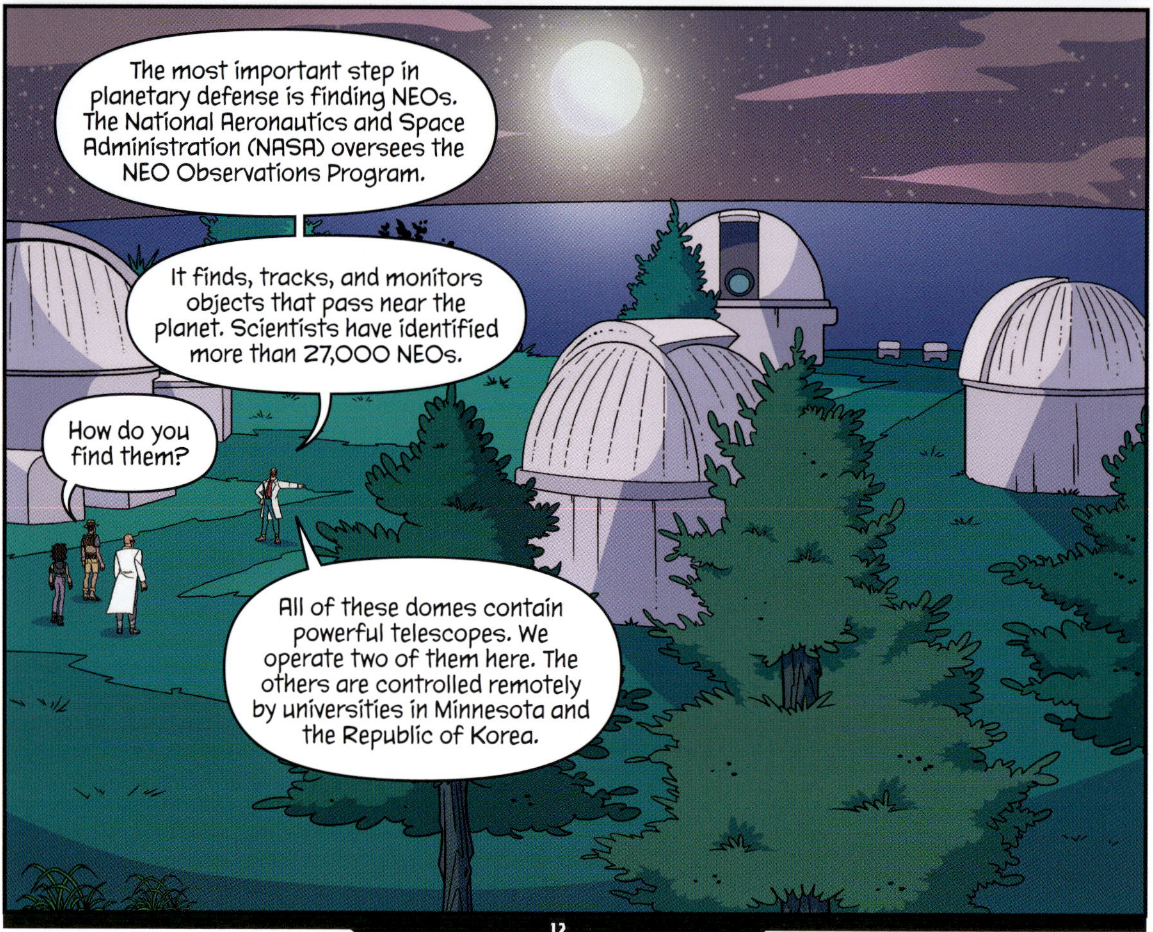
The most important step in planetary defense is finding NEOs. The National Aeronautics and Space Administration (NASA) oversees the NEO Observations Program.
It finds, tracks, and monitors objects that pass near the planet. Scientists have identified more than 27,000 NEOs.
How do you find them?
All of these domes contain powerful telescopes. We operate two of them here. The others are controlled remotely by universities in Minnesota and the Republic of Korea.

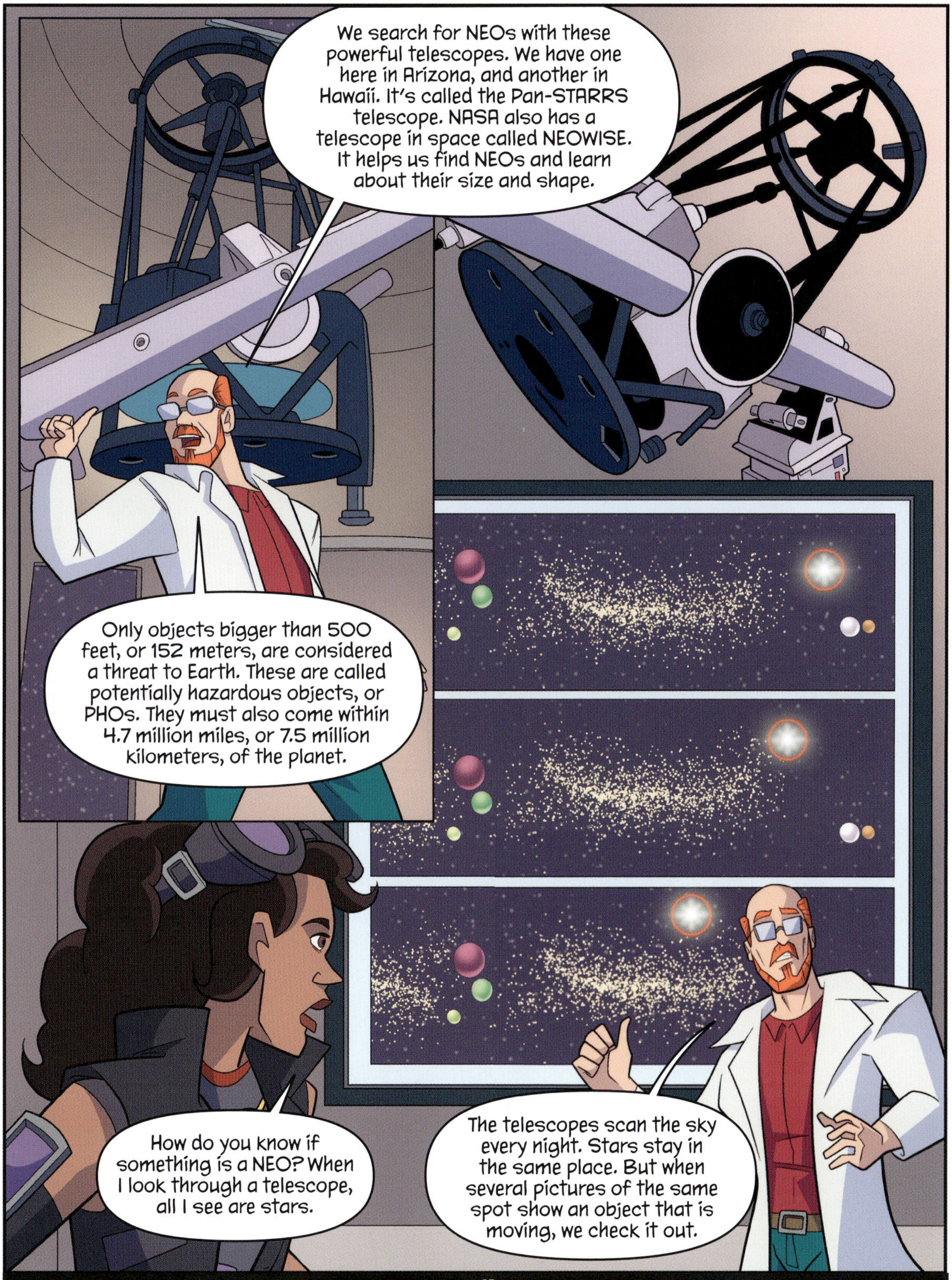
We search for NEOs with these powerful telescopes. We have one here in Arizona, and another in Hawaii. It's called the Pan-STARRS telescope. NASA also has a telescope in space called NEOWISE. It helps us find NEOs and learn about their size and shape.
Only objects bigger than 500 feet, or 152 meters, are considered a threat to Earth. These are called potentially hazardous objects, or PHOs. They must also come within 4.7 million miles, or 7.5 million kilometers, of the planet.
How do you know if something is a NEO? When I look through a telescope, all I see are stars.
The telescopes scan the sky every night. Stars stay in the same place. But when several pictures of the same spot show an object that is moving, we check it out.

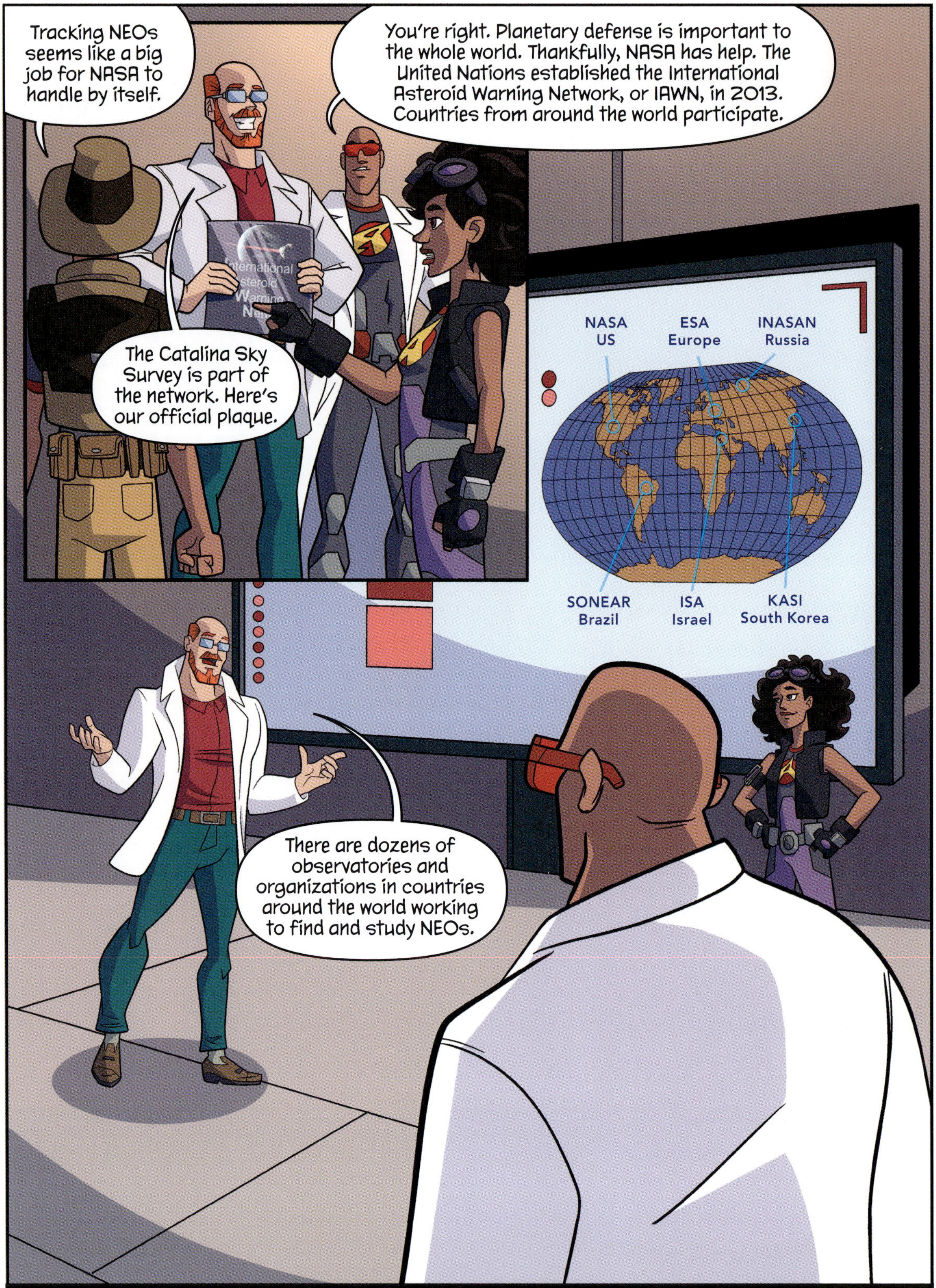
Tracking NEOs seems like a big job for NASA to handle by itself.
You're right. Planetary defense is important to the whole world. Thankfully, NASA has help. The United Nations established the International Asteroid Warning Network, or IAWN, in 2013. Countries from around the world participate.
The Catalina Sky Survey is part of the network. Here's our official plaque.
NASA US
ESA Europe
INASAN Russia
SONEAR Brazil
ISA Israel
KASI South Korea
There are dozens of observatories and organizations in countries around the world working to find and study NEOs.

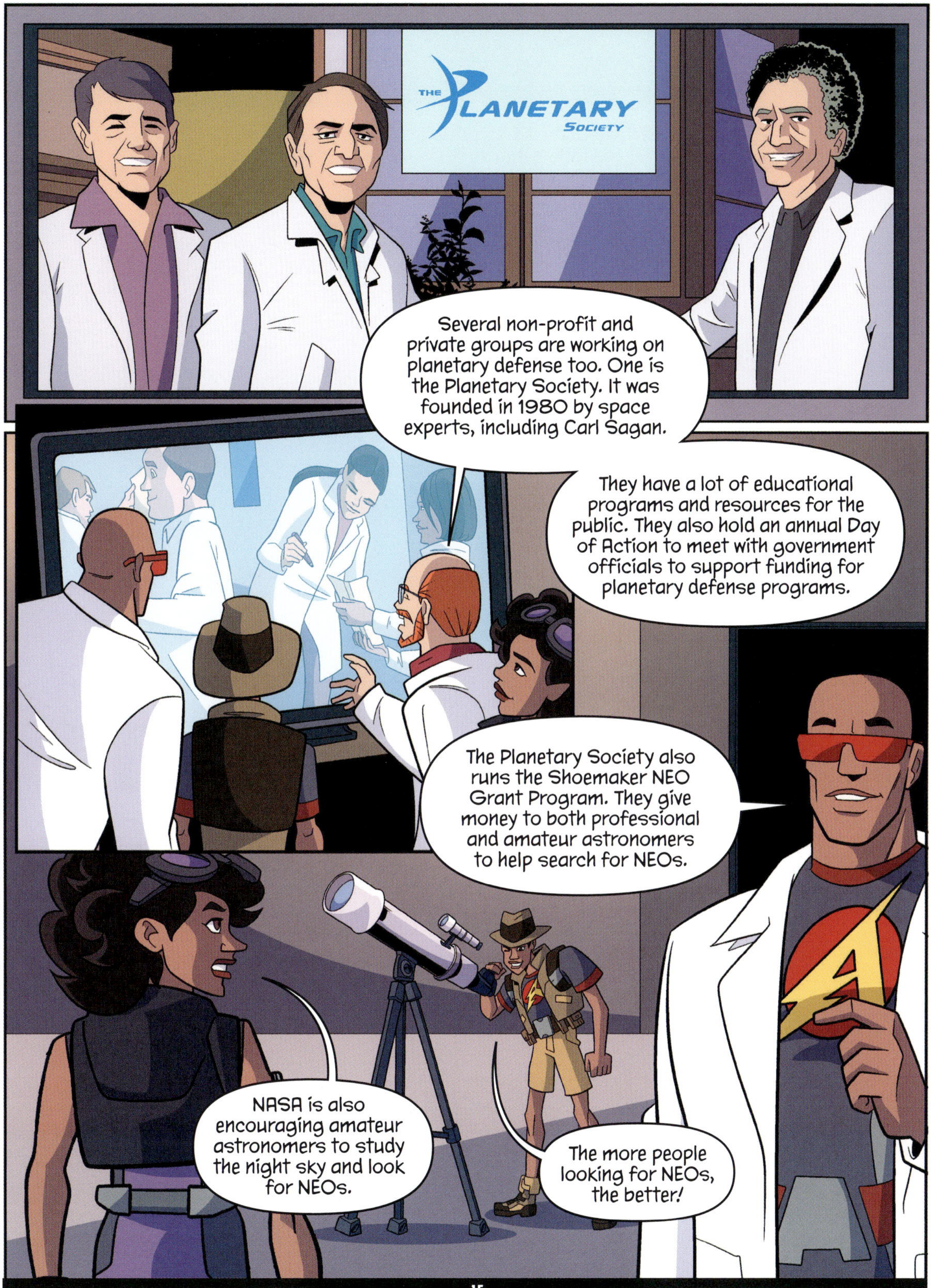
THE PLANETARY SOCIETY
Several non-profit and private groups are working on planetary defense too. One is the Planetary Society. It was founded in 1980 by space experts, including Carl Sagan.
They have a lot of educational programs and resources for the public. They also hold an annual Day of Action to meet with government officials to support funding for planetary defense programs.
The Planetary Society also runs the Shoemaker NEO Grant Program. They give money to both professional and amateur astronomers to help search for NEOs.
NASA is also encouraging amateur astronomers to study the night sky and look for NEOs.
The more people looking for NEOs, the better!

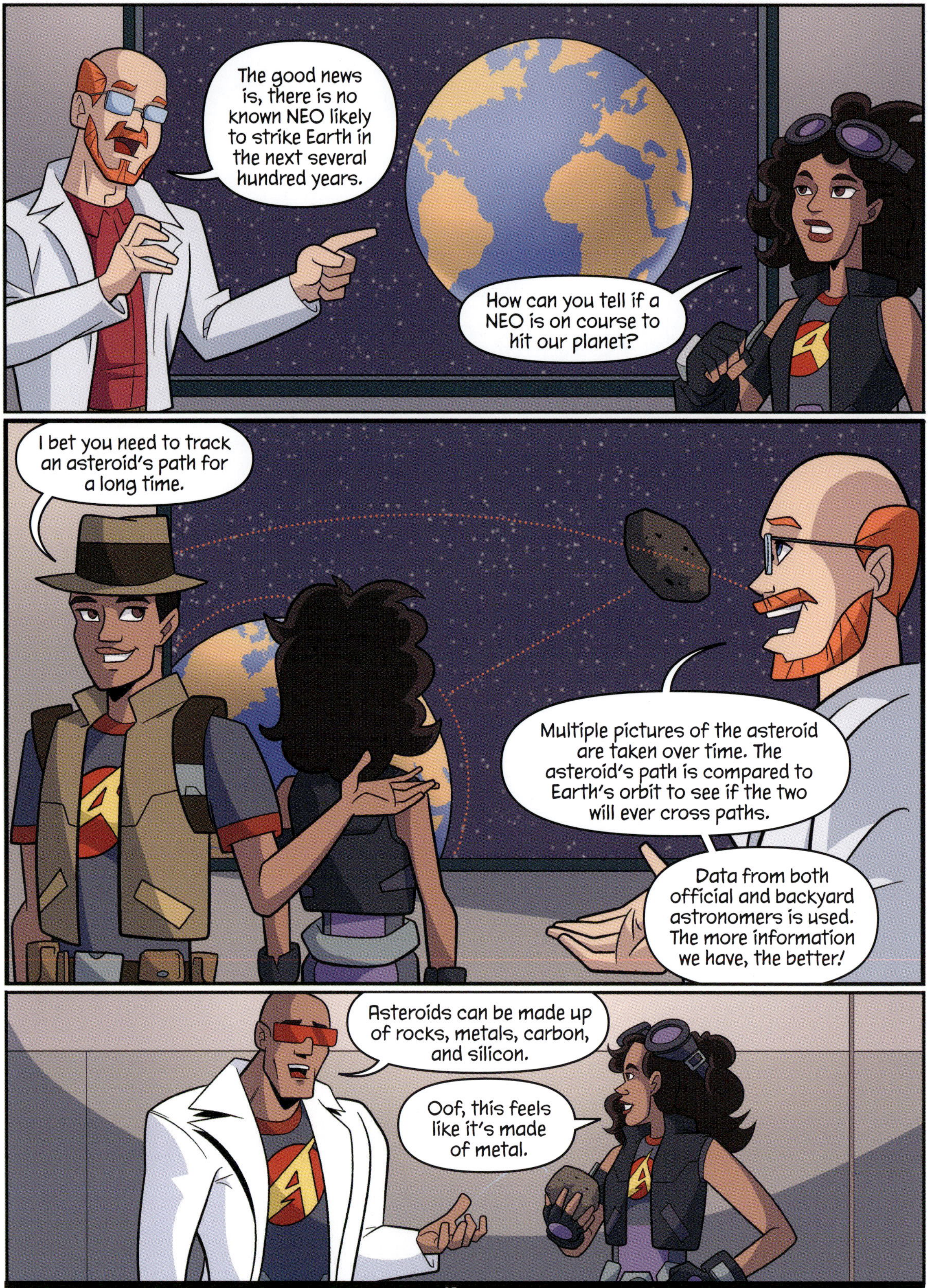
The good news is, there is no known NEO likely to strike Earth in the next several hundred years.
How can you tell if a NEO is on course to hit our planet?
I bet you need to track an asteroid's path for a long time.
Multiple pictures of the asteroid are taken over time. The asteroid's path is compared to Earth's orbit to see if the two will ever cross paths.
Data from both official and backyard astronomers is used. The more information we have, the better!
Asteroids can be made up of rocks, metals, carbon, and silicon.
Oof, this feels like it's made of metal.

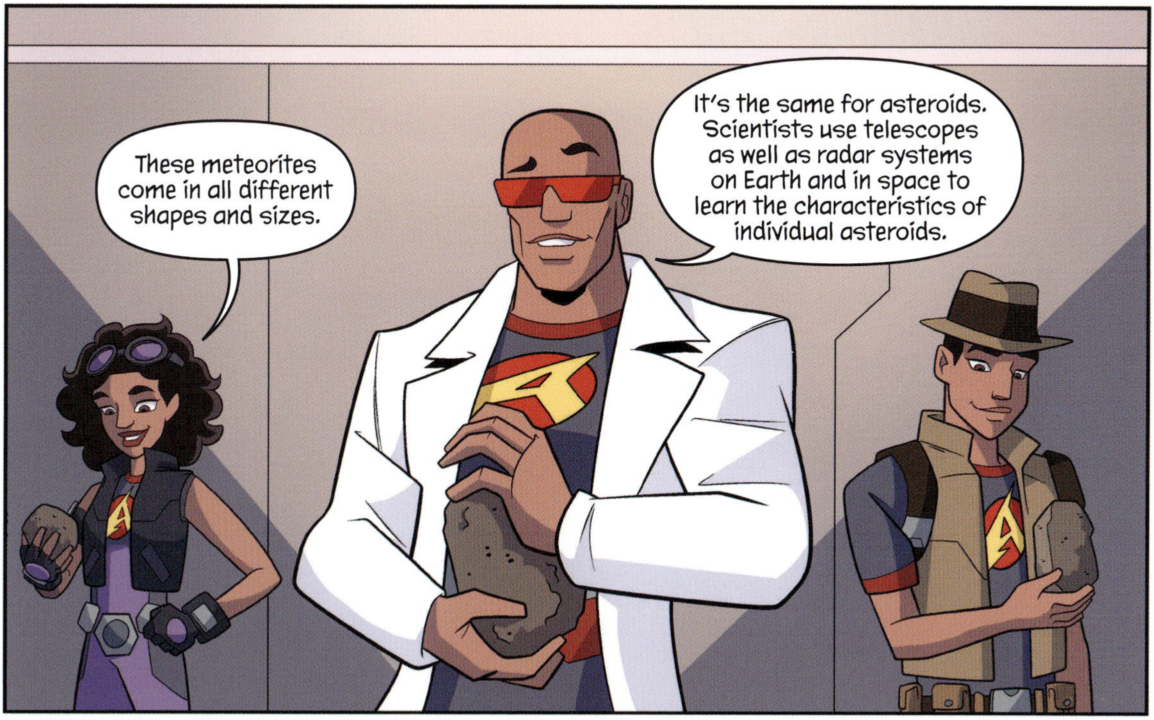
These meteorites come in all different shapes and sizes.
It's the same for asteroids. Scientists use telescopes as well as radar systems on Earth and in space to learn the characteristics of individual asteroids.

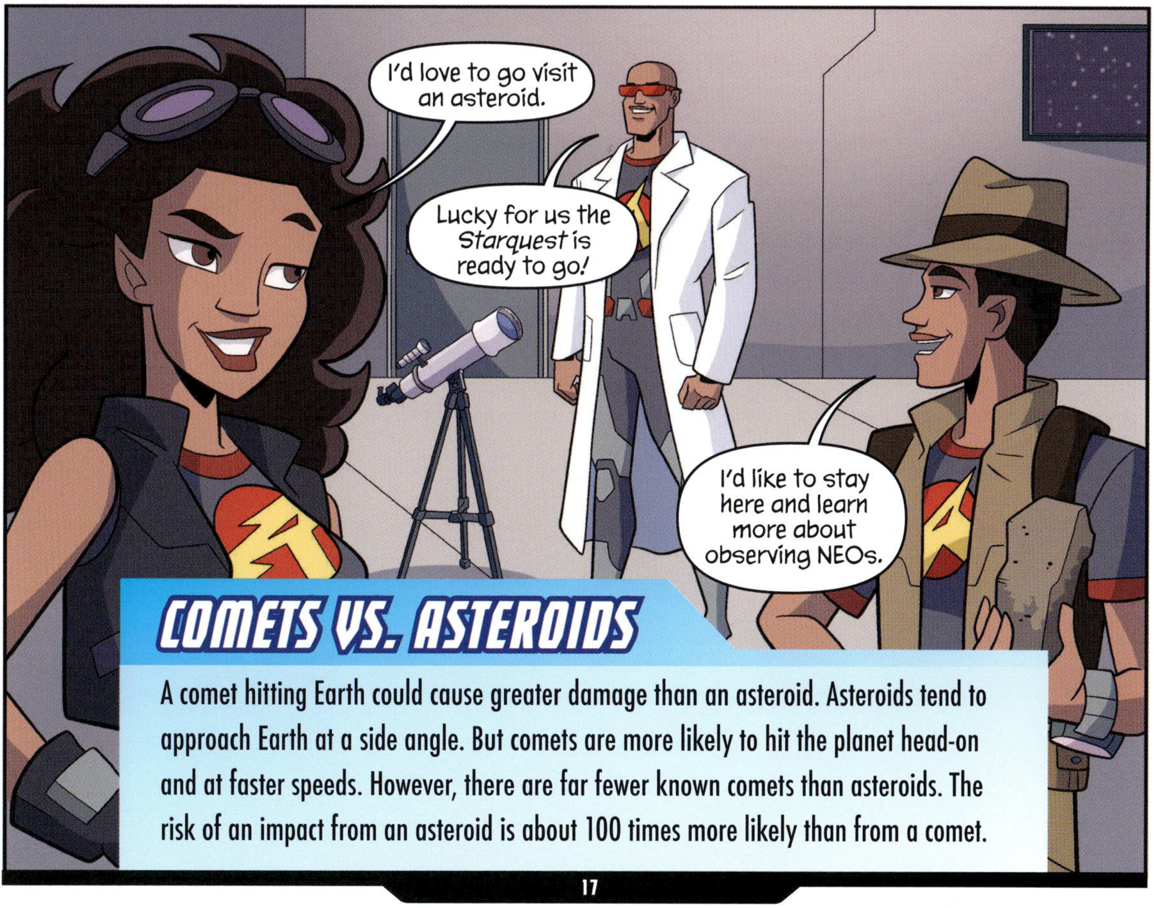
I'd love to go visit an asteroid.
Lucky for us the *Starquest* is ready to go!
I'd like to stay here and learn more about observing NEOs.

## COMETS VS. ASTEROIDS

A comet hitting Earth could cause greater damage than an asteroid. Asteroids tend to approach Earth at a side angle. But comets are more likely to hit the planet head-on and at faster speeds. However, there are far fewer known comets than asteroids. The risk of an impact from an asteroid is about 100 times more likely than from a comet.

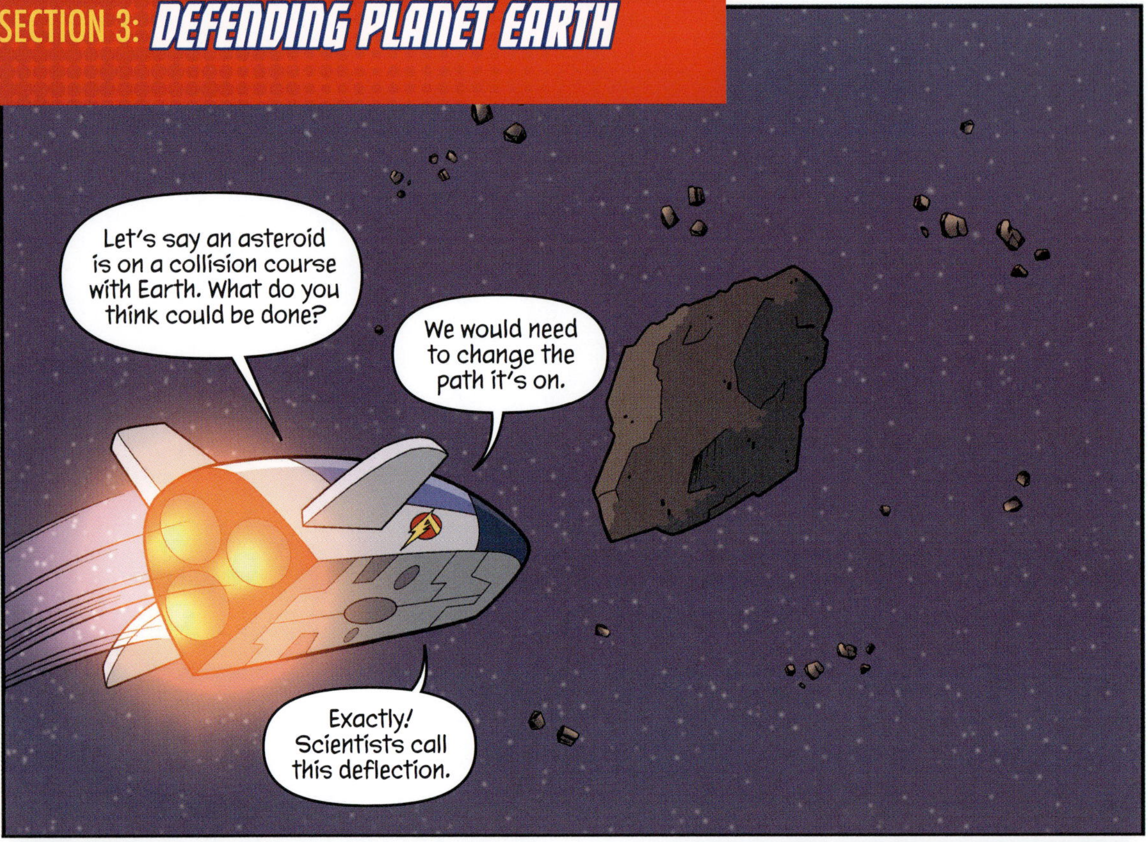
Let's say an asteroid is on a collision course with Earth. What do you think could be done?
We would need to change the path it's on.
Exactly! Scientists call this deflection.

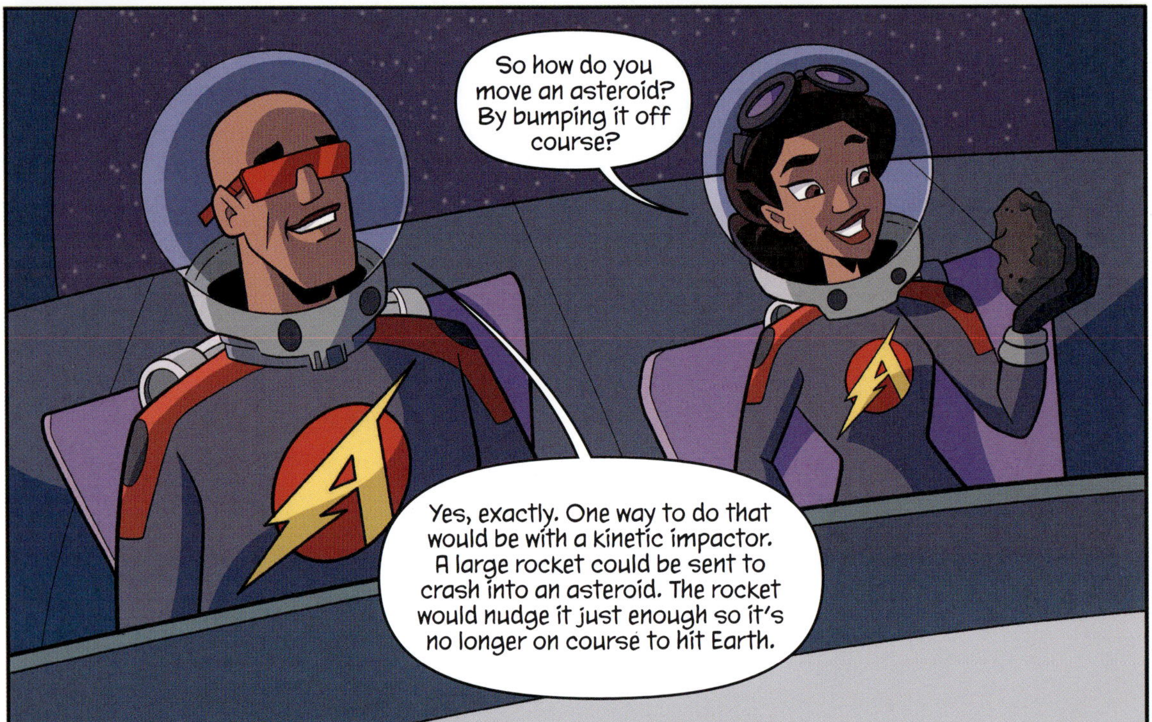
So how do you move an asteroid? By bumping it off course?
Yes, exactly. One way to do that would be with a kinetic impactor. A large rocket could be sent to crash into an asteroid. The rocket would nudge it just enough so it's no longer on course to hit Earth.

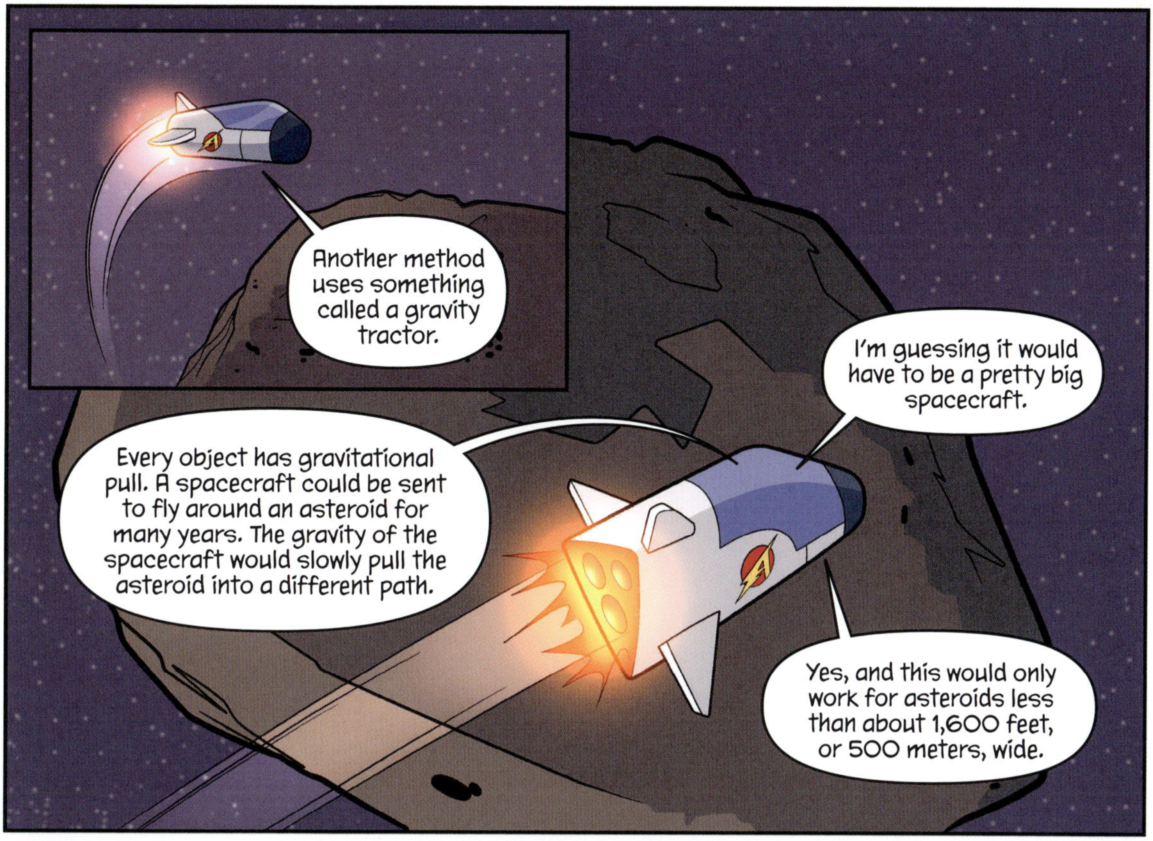
Another method uses something called a gravity tractor.
Every object has gravitational pull. A spacecraft could be sent to fly around an asteroid for many years. The gravity of the spacecraft would slowly pull the asteroid into a different path.
I'm guessing it would have to be a pretty big spacecraft.
Yes, and this would only work for asteroids less than about 1,600 feet, or 500 meters, wide.

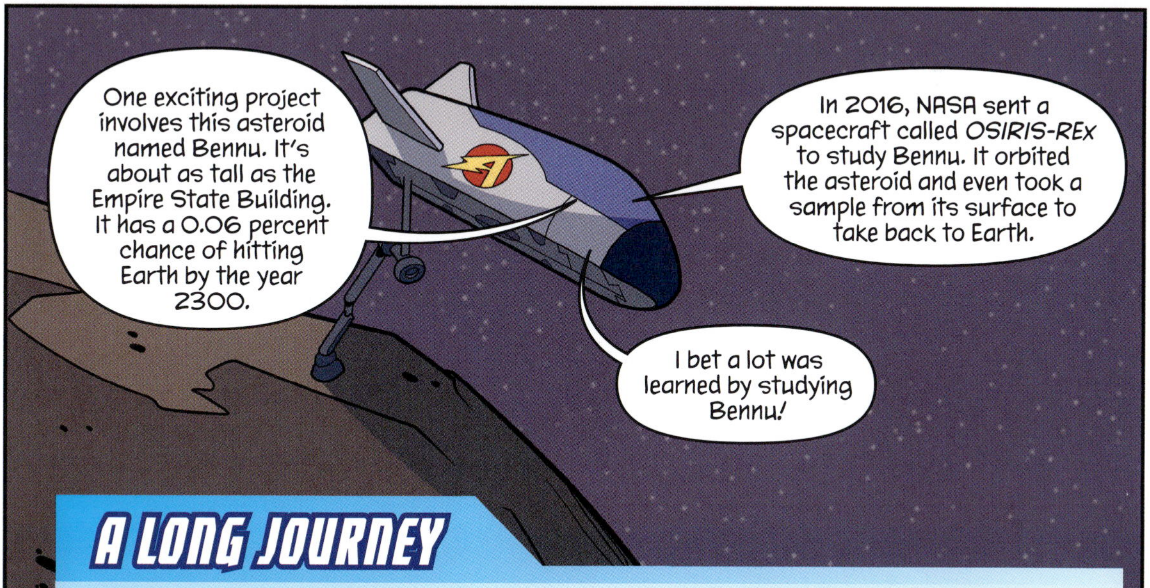
One exciting project involves this asteroid named Bennu. It's about as tall as the Empire State Building. It has a 0.06 percent chance of hitting Earth by the year 2300.
In 2016, NASA sent a spacecraft called *OSIRIS-REx* to study Bennu. It orbited the asteroid and even took a sample from its surface to take back to Earth.
I bet a lot was learned by studying Bennu!

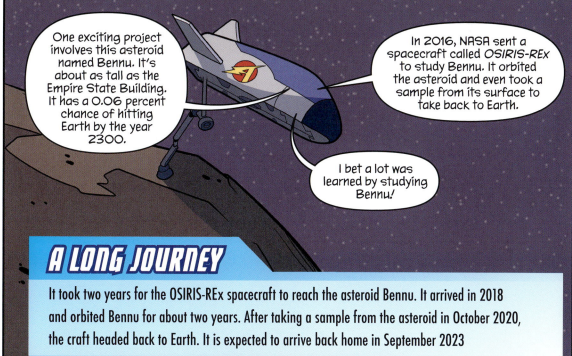

## A LONG JOURNEY

It took two years for the OSIRIS-REx spacecraft to reach the asteroid Bennu. It arrived in 2018 and orbited Bennu for about two years. After taking a sample from the asteroid in October 2020, the craft headed back to Earth. It is expected to arrive back home in September 2023

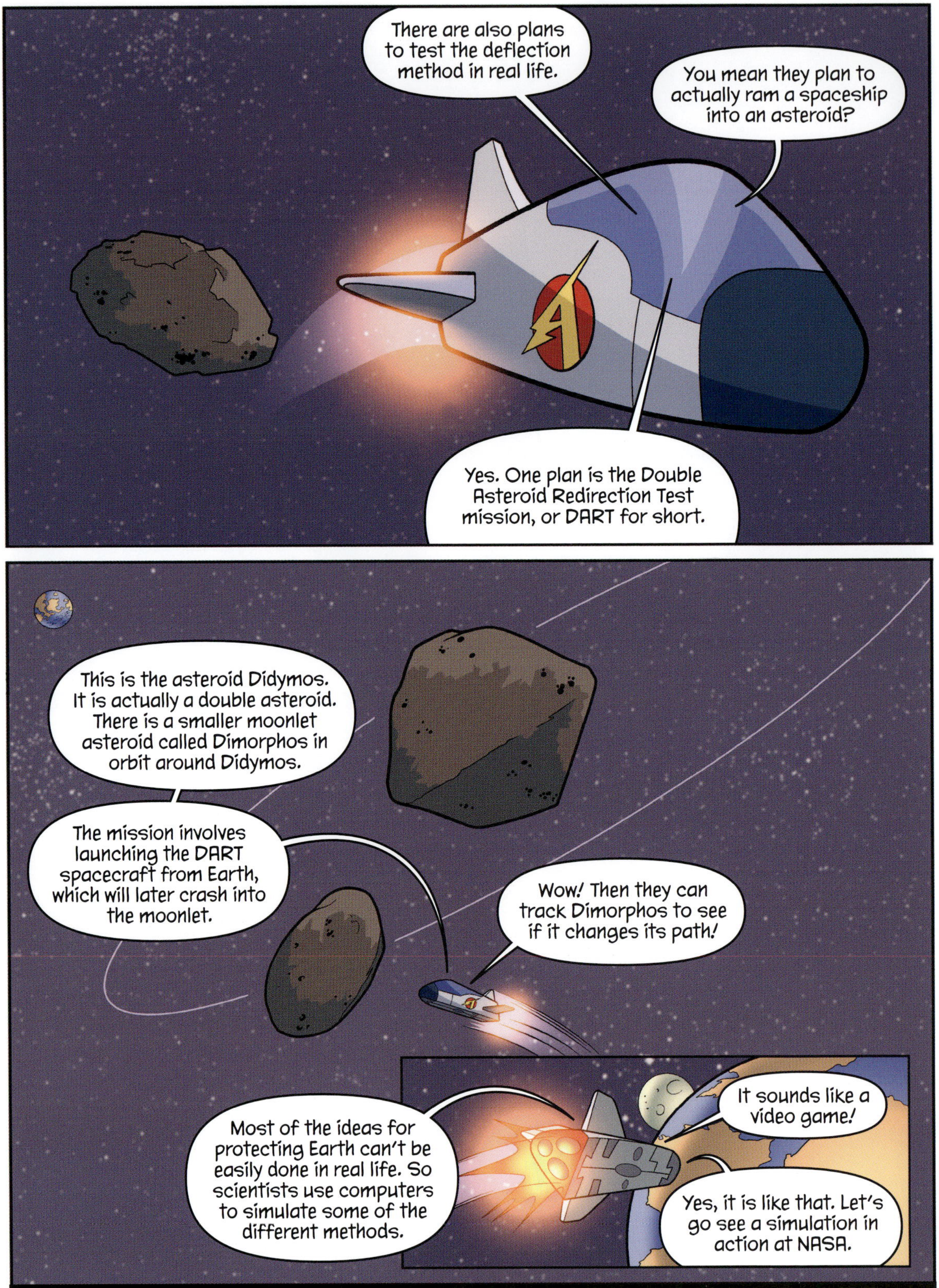
There are also plans to test the deflection method in real life.
You mean they plan to actually ram a spaceship into an asteroid?
Yes. One plan is the Double Asteroid Redirection Test mission, or DART for short.
This is the asteroid Didymos. It is actually a double asteroid. There is a smaller moonlet asteroid called Dimorphos in orbit around Didymos.
The mission involves launching the DART spacecraft from Earth, which will later crash into the moonlet.
Wow! Then they can track Dimorphos to see if it changes its path!
Most of the ideas for protecting Earth can't be easily done in real life. So scientists use computers to simulate some of the different methods.
It sounds like a video game!
Yes, it is like that. Let's go see a simulation in action at NASA.

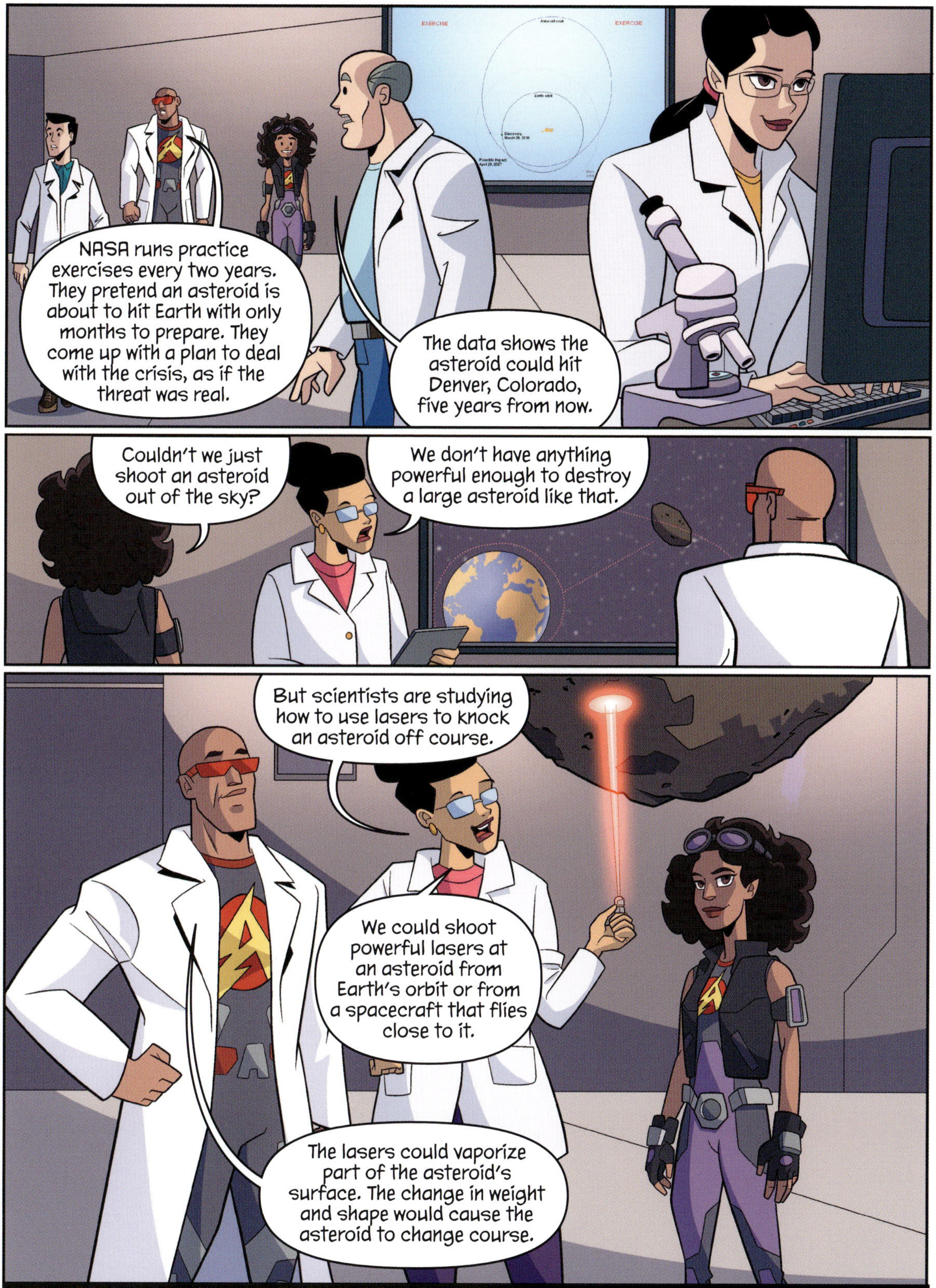
NASA runs practice exercises every two years. They pretend an asteroid is about to hit Earth with only months to prepare. They come up with a plan to deal with the crisis, as if the threat was real.
The data shows the asteroid could hit Denver, Colorado, five years from now.
Couldn't we just shoot an asteroid out of the sky?
We don't have anything powerful enough to destroy a large asteroid like that.
But scientists are studying how to use lasers to knock an asteroid off course.
We could shoot powerful lasers at an asteroid from Earth's orbit or from a spacecraft that flies close to it.
The lasers could vaporize part of the asteroid's surface. The change in weight and shape would cause the asteroid to change course.

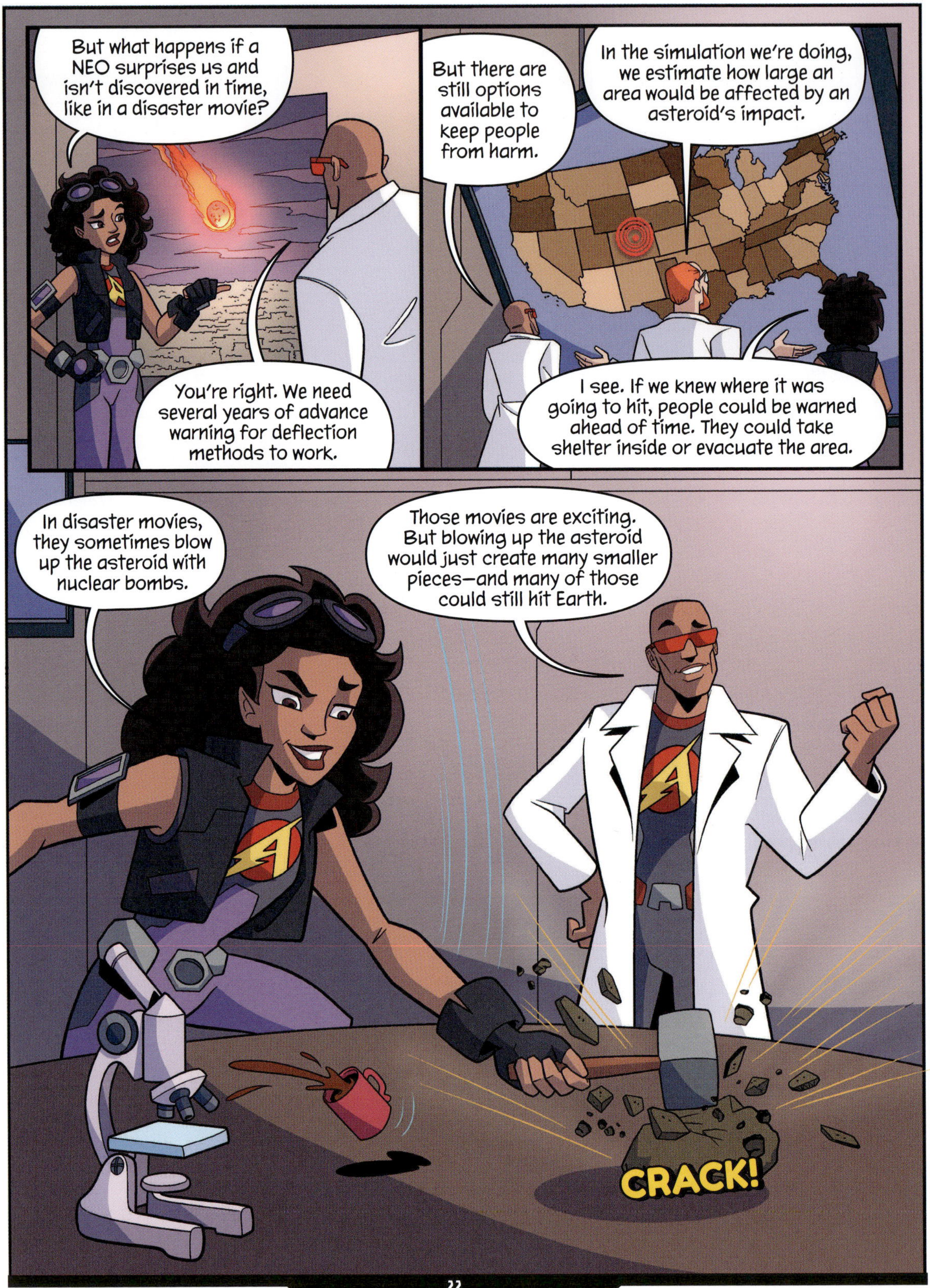
But what happens if a NEO surprises us and isn't discovered in time, like in a disaster movie?
You're right. We need several years of advance warning for deflection methods to work.
But there are still options available to keep people from harm.
In the simulation we're doing, we estimate how large an area would be affected by an asteroid's impact.
I see. If we knew where it was going to hit, people could be warned ahead of time. They could take shelter inside or evacuate the area.
In disaster movies, they sometimes blow up the asteroid with nuclear bombs.
Those movies are exciting. But blowing up the asteroid would just create many smaller pieces—and many of those could still hit Earth.
CRACK!

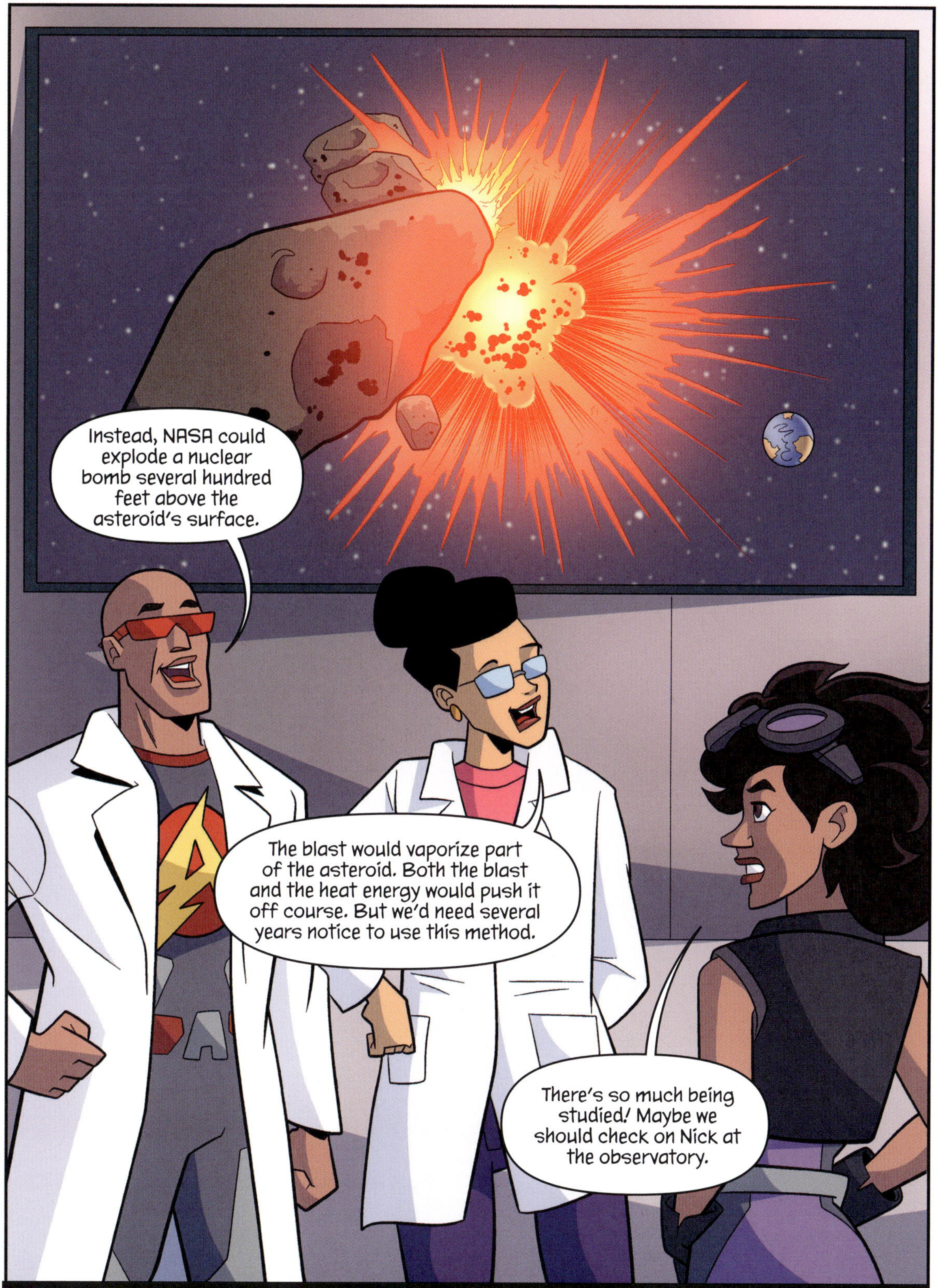
Instead, NASA could explode a nuclear bomb several hundred feet above the asteroid's surface.
The blast would vaporize part of the asteroid. Both the blast and the heat energy would push it off course. But we'd need several years notice to use this method.
There's so much being studied! Maybe we should check on Nick at the observatory.

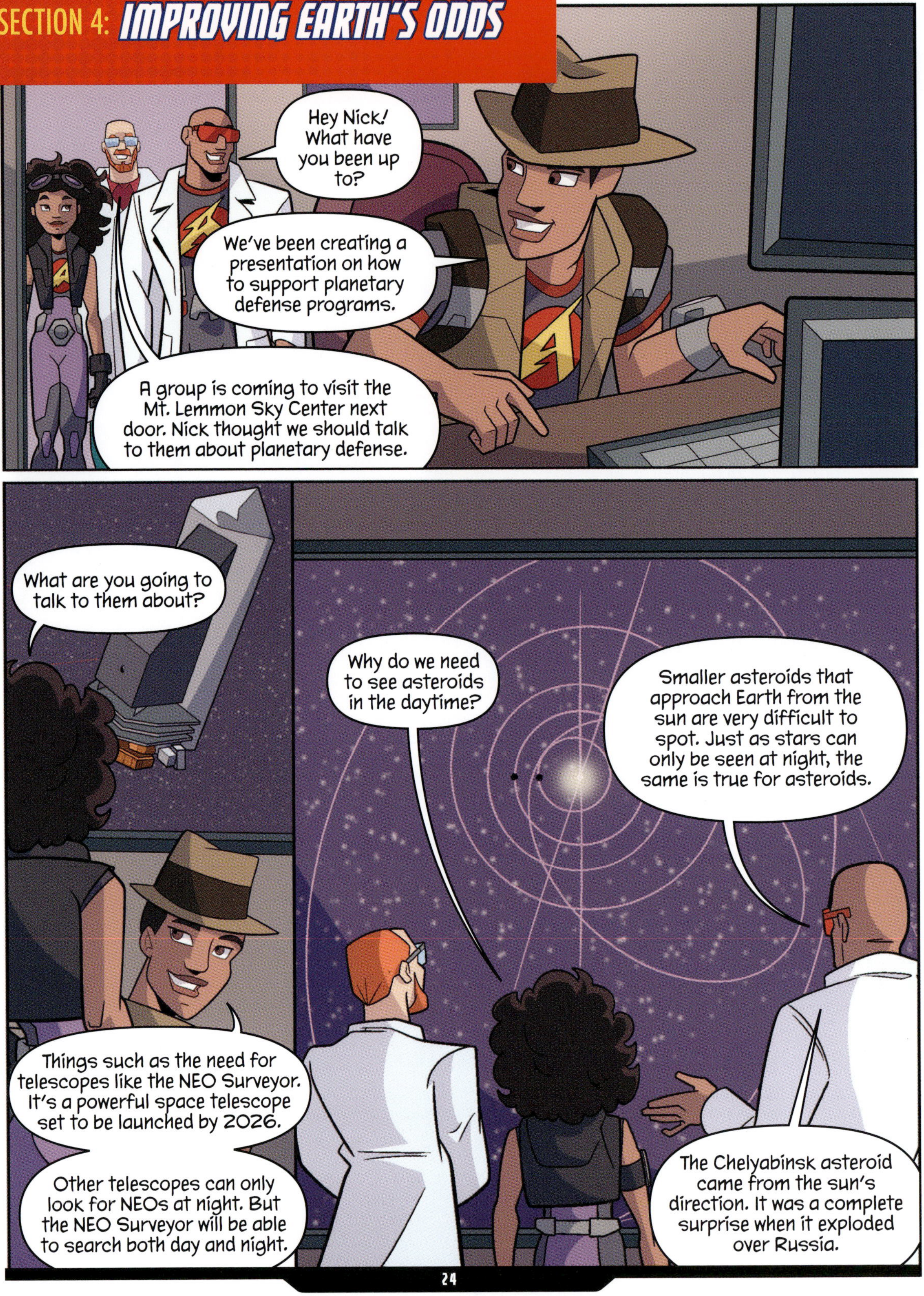
Hey Nick! What have you been up to?
We've been creating a presentation on how to support planetary defense programs.
A group is coming to visit the Mt. Lemmon Sky Center next door. Nick thought we should talk to them about planetary defense.
What are you going to talk to them about?
Things such as the need for telescopes like the NEO Surveyor. It's a powerful space telescope set to be launched by 2026.
Other telescopes can only look for NEOs at night. But the NEO Surveyor will be able to search both day and night.
Why do we need to see asteroids in the daytime?
Smaller asteroids that approach Earth from the sun are very difficult to spot. Just as stars can only be seen at night, the same is true for asteroids.
The Chelyabinsk asteroid came from the sun's direction. It was a complete surprise when it exploded over Russia.

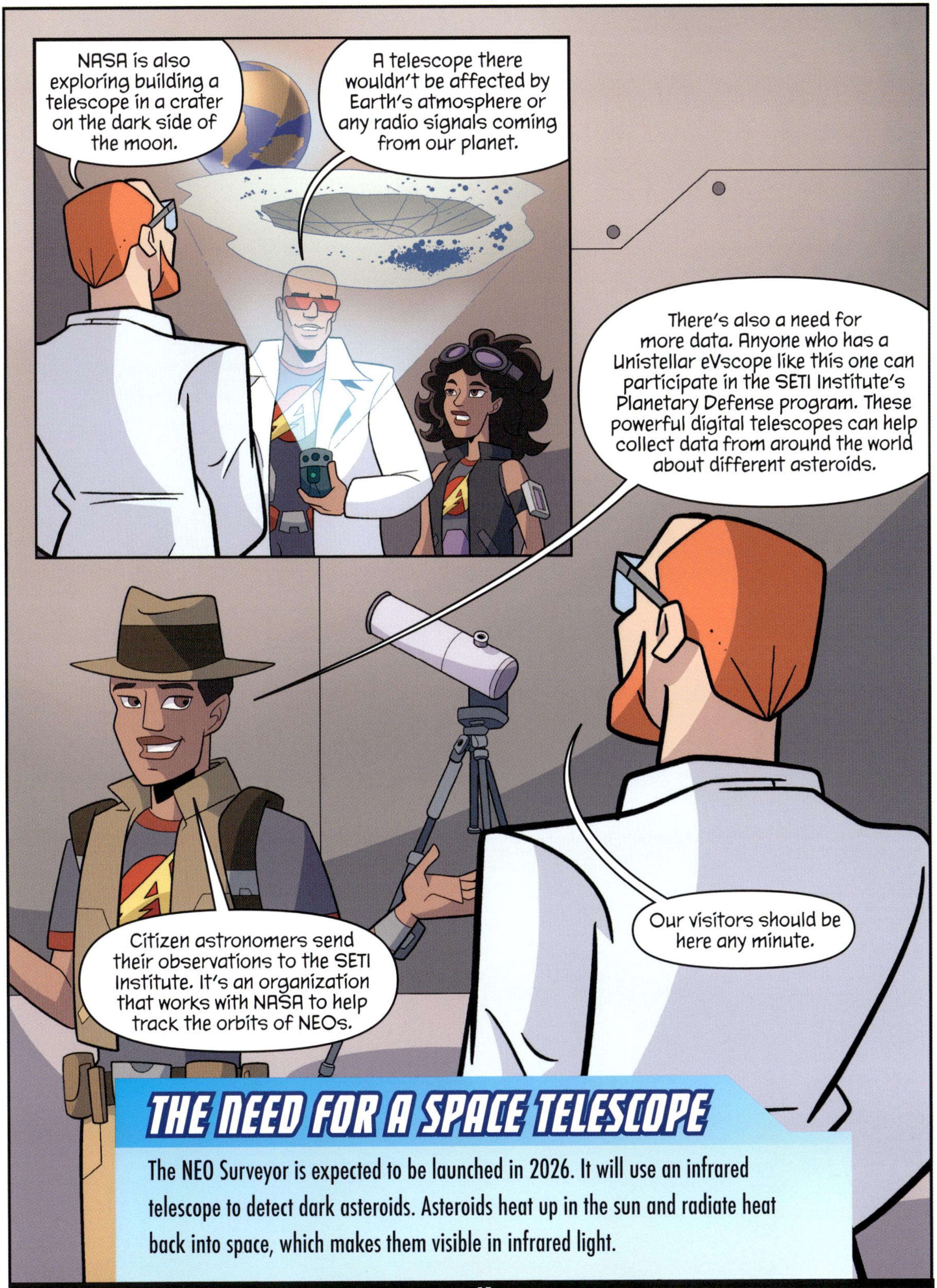
NASA is also exploring building a telescope in a crater on the dark side of the moon.
A telescope there wouldn't be affected by Earth's atmosphere or any radio signals coming from our planet.
There's also a need for more data. Anyone who has a Unistellar eVscope like this one can participate in the SETI Institute's Planetary Defense program. These powerful digital telescopes can help collect data from around the world about different asteroids.
Citizen astronomers send their observations to the SETI Institute. It's an organization that works with NASA to help track the orbits of NEOs.
Our visitors should be here any minute.

## THE NEED FOR A SPACE TELESCOPE

The NEO Surveyor is expected to be launched in 2026. It will use an infrared telescope to detect dark asteroids. Asteroids heat up in the sun and radiate heat back into space, which makes them visible in infrared light.

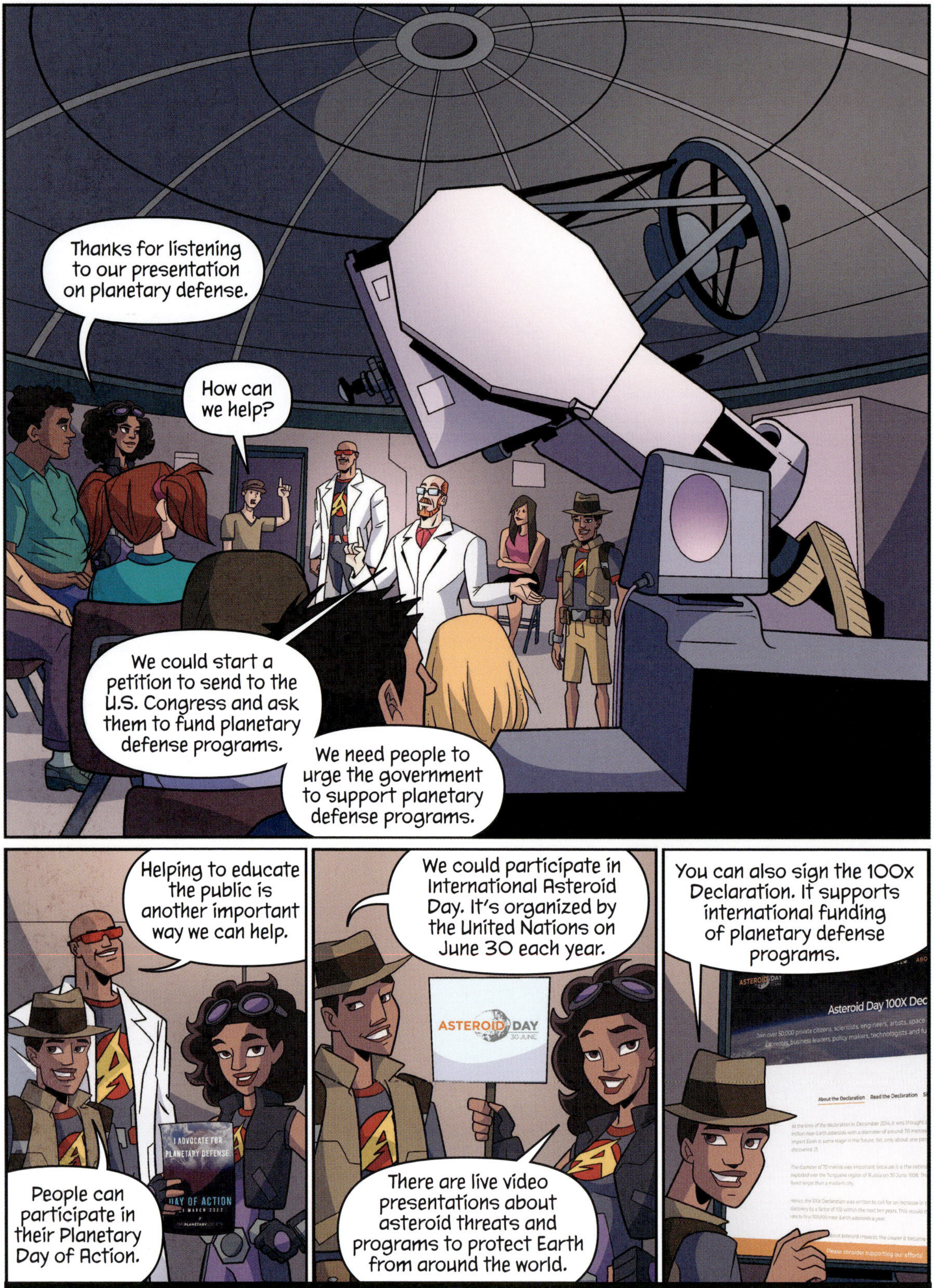
Thanks for listening to our presentation on planetary defense.
How can we help?
We could start a petition to send to the U.S. Congress and ask them to fund planetary defense programs.
We need people to urge the government to support planetary defense programs.
Helping to educate the public is another important way we can help.
I ADVOCATE FOR PLANETARY DEFENSE
DAY OF ACTION
People can participate in their Planetary Day of Action.
We could participate in International Asteroid Day. It's organized by the United Nations on June 30 each year.
ASTEROID DAY
30 JUNE
There are live video presentations about asteroid threats and programs to protect Earth from around the world.
You can also sign the 100x Declaration. It supports international funding of planetary defense programs.

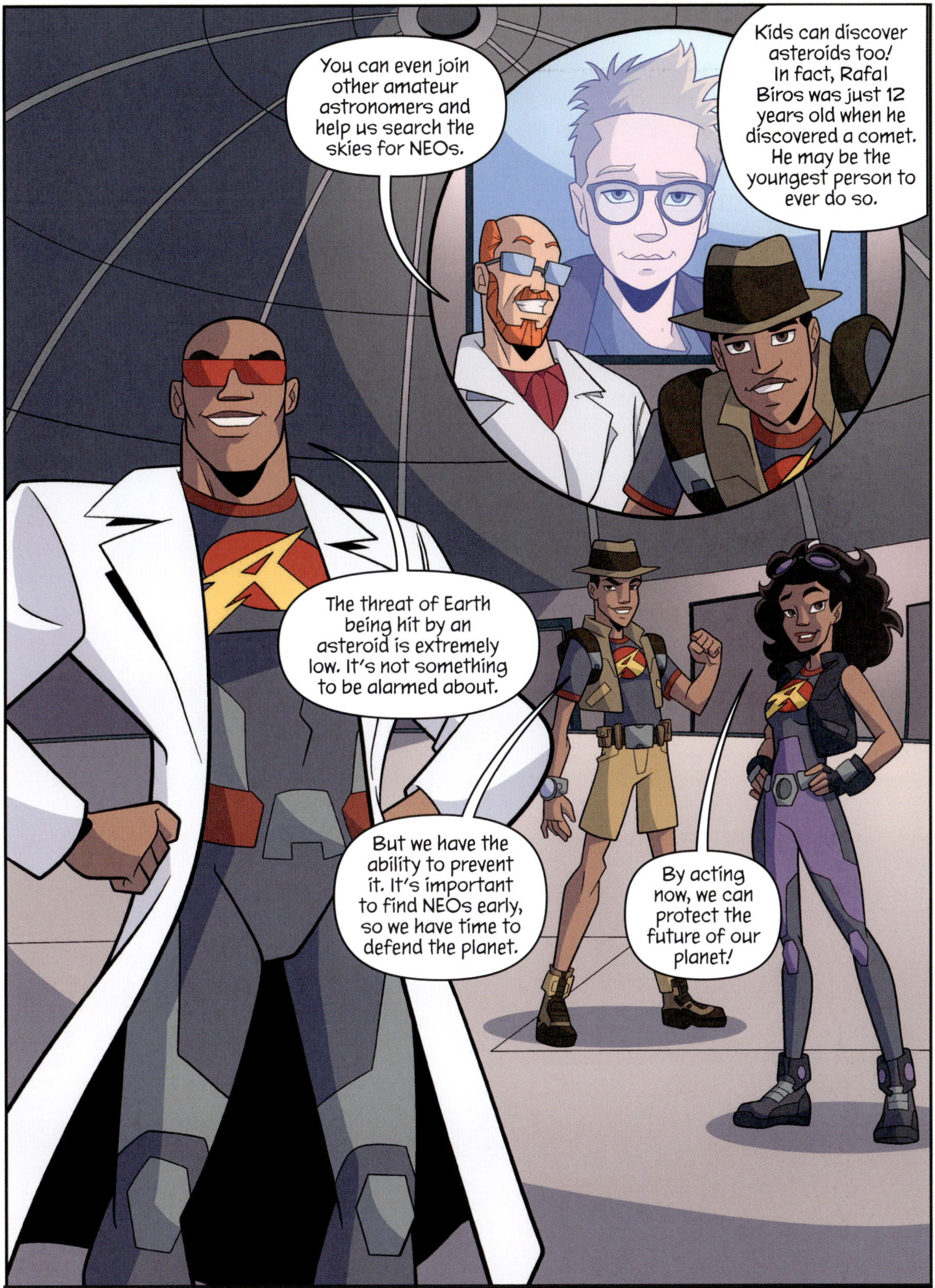
You can even join other amateur astronomers and help us search the skies for NEOs.
Kids can discover asteroids too! In fact, Rafal Biros was just 12 years old when he discovered a comet. He may be the youngest person to ever do so.
The threat of Earth being hit by an asteroid is extremely low. It's not something to be alarmed about.
But we have the ability to prevent it. It's important to find NEOs early, so we have time to defend the planet.
By acting now, we can protect the future of our planet!

# MORE ABOUT NEAR EARTH OBJECTS

Until 2021, it was believed the asteroid Apophis had a chance of hitting Earth in 2029. However, scientists have determined it won't be a threat for at least another 100 years. But it will pass near Earth on Friday, April 13, 2029. Apophis will pass by at a distance of only 19,662 miles (31,643 km.)

The first NEO to be discovered was Eros in 1898. It is the second largest known NEO at 10.5 miles (16.8 km) in length. It is also the first asteroid to have a spacecraft land on it. In February 2001, the *NEAR-Shoemaker* spacecraft unexpectedly survived the landing and for two weeks sent images from the asteroid back to Earth.

Asteroids orbit the sun just like the planets. But their orbits can change when they're nudged by the gravitational pull of the planets.

In 1994, fragments from Comet Shoemaker-Levy 9 crashed into Jupiter. It broke into about 20 large pieces and caused a series of explosions. The explosions were as strong as about 10 million megatons of dynamite.

In 2020, 107 known asteroids passed closer to Earth than the moon. The moon is 240,000 miles (386,000 km) from Earth.

The asteroid belt is huge. But even though there are hundreds of thousands of asteroids in the belt, it isn't very crowded. There is more than 600,000 miles (966,000 km) of space between asteroids. Many spacecraft have easily traveled through the asteroid belt without running into any asteroids.

In 2016 a team of scientists drilled into the crater left by the asteroid believed to have wiped out the dinosaurs. By studying the core sample, they found high levels of iridium. This element is commonly found in and near impact craters. The discovery left little doubt that the crater was in fact the site of the crash that led to the end of the dinosaurs.

In July 2020, Comet NEOWISE provided the world with a rare opportunity to view a bright comet in the sky. Its existence surprised NASA scientists. It was discovered almost by accident by the NEOWISE telescope in March 2020. It was found just three months before it made its close approach to Earth. Fortunately, it was not on a collision course with our planet.

# GLOSSARY

**amateur** (AM-uh-chur)—someone who does something for fun rather than money

**atmosphere** (AT-muh-sfeer)—the layer of gases that surrounds some planets, dwarf planets, and moons

**crater** (KRAY-tuhr)—a large hole made when large pieces of rock crash into a planet or moon's surface

**deflection** (dih-FLEK-shuhn)—an action that causes an object to travel in a different direction

**kinetic** (kih-NET-ik)—having to do with motion

**laser** (LAY-zur)—a thin, intense beam of light

**observatory** (uhb-ZUR-vuh-tohr-ee)—a building that uses telescopes or other instruments to study the stars

**orbit** (OR-bit)—the path an object follows as it goes around the Sun or a planet

**petition** (puh-TISH-uhn)—a letter signed by many people asking leaders for a change

**radar** (RAY-dahr)—a device that uses radio waves to track the location, distance, and speed of objects

**vaporize** (VAY-puh-rahyz)—to convert something into a vapor or gas

## READ MORE

Kurtz, Kevin. *Comets and Asteroids in Action: An Augmented Reality Experience.* Minneapolis: Lerner Publications, 2020.

Labrecque, Ellen. *Mysteries of Meteors, Asteroids, and Comets.* North Mankato, MN: Capstone, 2021.

London, Martha. *Explore Asteroids.* Minneapolis: Abdo Publishing, 2021.

Owings, Lisa. *Asteroid Impact.* Minneapolis: Bellwether Media, 2020.

## INTERNET SITES

*Astronomy for Kids: Asteroids*
ducksters.com/science/physics/asteroids.php

*The Planetary Society: Asteroid Defense 101*
courses.planetary.org/p/asteroid-defense-101

*NASA Science: Space Place—Asteroid*
spaceplace.nasa.gov/menu/asteroid/

*NEO Basics*
cneos.jpl.nasa.gov/about/basics.html

# INDEX

asteroids
  Apophis, 28
  asteroid belt, 8, 29
  Bennu, 19
  Didymos and Dimorphos, 20
  Eros, 28
  impact events, 10, 24, 29

Biros, Rafal, 27

comets, 8, 17, 27
  Comet NEOWISE, 6–7, 29
  Comet Shoemaker-Levy 9, 11, 28

Double Asteroid Redirection Test (DART), 20

International Asteroid Warning Network (IAWN), 14

meteors and meteorites, 8, 9, 17

NASA, 12, 13, 14, 15, 21, 23, 25, 29
near-Earth objects (NEOs), 7, 8, 9, 11, 12, 14, 15, 22, 24, 25, 27
  potentially hazardous objects (PHOs), 13
  tracking of, 16

NEAR-Shoemaker, 28
NEO Observations Program, 12

OSIRIS-REx, 19

Planetary Society, 15
protecting Earth, 27
  100x Declaration, 26
  deflection method, 18, 20, 22
  gravity tractor method, 19
  International Asteroid Day, 26
  lasers, 21
  nuclear bombs, 22–23
  Planetary Day of Action, 26
  simulations, 20–22

Sagan, Carl, 15
SETI Institute, 25
Steward Observatory, 12–13

telescopes, 12–13, 17, 25
  NEO Surveyor, 24, 25
  NEOWISE, 13, 29
  Pan-STARRS telescope, 13

# ABOUT THE AUTHOR

Carol Kim is the author of several fiction and nonfiction books for kids. She enjoys researching and uncovering little-known facts and sharing what she learns with young readers. Carol lives in Austin, Texas, with her family. Learn more about her and her latest books at her website, CarolKimBooks.com.